Solving Mathematical Problems

Solving Mathematical Problems

A Personal Perspective

Terence Tao
Department of Mathematics, UCLA, Los Angeles, CA 90095

UNIVERSITY PRESS

Great Clarendon Street, Oxford OX2 6DP
Oxford University Press is a department of the University of Oxford.
It furthers the University's objective of excellence in research, scholarship,
and education by publishing worldwide in
Oxford New York
Auckland   Cape Town   Dar es Salaam   Hong Kong   Karachi
Kuala Lumpur   Madrid   Melbourne   Mexico City   Nairobi
New Delhi   Shanghai   Taipei   Toronto
With offices in
Argentina   Austria   Brazil   Chile   Czech Republic   France   Greece
Guatemala   Hungary   Italy   Japan   South Korea   Poland   Portugal
Singapore   Switzerland   Thailand   Turkey   Ukraine   Vietnam

Oxford is a registered trade mark of Oxford University Press
in the UK and in certain other countries

Published in the United States
by Oxford University Press Inc., New York

© Terence Tao, 2006

The moral rights of the author have been asserted

Database right Oxford University Press (maker)

Reprinted 2010

All rights reserved. No part of this publication may be reproduced,
stored in a retrieval system, or transmitted, in any form or by any means,
without the prior permission in writing of Oxford University Press,
or as expressly permitted by law, or under terms agreed with the appropriate
reprographics rights organization. Enquiries concerning reproduction
outside the scope of the above should be sent to the Rights Department,
Oxford University Press, at the address above

You must not circulate this book in any other binding or cover
And you must impose this same condition on any acquirer

ISBN 978-0-19-920560-8

Printed in the United Kingdom by
Lightning Source UK Ltd., Milton Keynes

Dedicated to all my mentors, who taught me the meaning (and joy) of mathematics.

Contents

Preface to the first edition viii

Preface to the second edition xi

1 Strategies in problem solving 1
2 Examples in number theory 9
3 Examples in algebra and analysis 35
4 Euclidean geometry 49
5 Analytic geometry 69
6 Sundry examples 83

References 99
Index 101

Preface to the first edition

Proclus, an ancient Greek philosopher, said:
> This therefore, is mathematics: she reminds you of the invisible forms of the soul; she gives life to her own discoveries; she awakens the mind and purifies the intellect; she brings to light our intrinsic ideas; she abolishes oblivion and ignorance which are ours by birth ...

But I just like mathematics because it is fun.

Mathematical problems, or puzzles, are important to real mathematics (like solving real-life problems), just as fables, stories, and anecdotes are important to the young in understanding real life. Mathematical problems are 'sanitized' mathematics, where an elegant solution has already been found (by someone else, of course), the question is stripped of all superfluousness and posed in an interesting and (hopefully) thought-provoking way. If mathematics is likened to prospecting for gold, solving a good mathematical problem is akin to a 'hide-and-seek' course in gold-prospecting: you are given a nugget to find, and you know what it looks like, that it is out there somewhere, that it is not too hard to reach, that it is unearthing within your capabilities, and you have conveniently been given the right equipment (i.e. data) to get it. It may be hidden in a cunning place, but it will require ingenuity rather than digging to reach it.

In this book I shall solve selected problems from various levels and branches of mathematics. Starred problems (*) indicate an additional level of difficulty, either because some higher mathematics or some clever thinking are required; double-starred questions (**) are similar, but to a greater degree. Some problems have additional exercises at the end that can be solved in a similar manner or involve a similar piece of mathematics. While solving these problems, I will try to demonstrate some tricks of the trade when problem-solving. Two of the main weapons—experience and knowledge—are not easy to put into a book: they have to be acquired over time. But there are many simpler tricks that take less time to learn. There are ways of looking at a problem that make it easier to find a feasible attack plan. There are systematic ways of reducing a problem into successively simpler sub-problems. But, on the other hand, solving the problem is not everything. To return to the gold nugget analogy, strip-mining the neighbourhood with bulldozers is clumsier than doing a careful survey, a bit of geology, and a small amount of digging. A solution should be relatively short, understandable, and hopefully have a touch of elegance. It should also be fun to discover. Transforming a nice, short little geometry question into a ravening monster of an equation by textbook coordinate geometry does not have the same taste of victory as a two-line vector solution.

As an example of elegance, here is a standard result in Euclidean geometry:

> Show that the perpendicular bisectors of a triangle are concurrent.

This neat little one-liner *could* be attacked by coordinate geometry. Try to do so for a few minutes (hours?), then look at this solution:

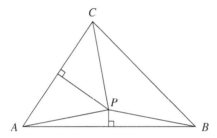

PROOF. Call the triangle ABC. Now let P be the intersection of the perpendicular bisectors of AB and AC. Because P is on the AB bisector, $|AP| = |PB|$. Because P is on the AC bisector, $|AP| = |PC|$. Combining the two, $|BP| = |PC|$. But this means that P has to be on the BC bisector. Hence all three bisectors are concurrent. (Incidentally, P is the circumcentre of ABC.) □

The following reduced diagram shows why $|AP| = |PB|$ if P is on the AB perpendicular bisector: congruent triangles will pull it off nicely.

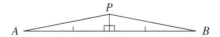

This kind of solution—and the strange way that obvious facts mesh to form a not-so-obvious fact—is part of the beauty of mathematics. I hope that you too will appreciate this beauty.

Acknowledgements

Thanks to Peter O'Halloran, Vern Treilibs, and Lenny Ng for their contributions of problems and advice.

Special thanks to Basil Rennie for his corrections and ingenious short-cuts in solutions, and finally thanks to my family for their support, encouragement, spelling corrections, and put-downs when I was behind schedule.

Almost all of the problems in this book come from published collections of problem sets for mathematics competitions. These are sourced in the texts, with full details given in the reference section of the book. I also used a small handful of problems from friends or from various mathematical publications; these have no source listed.

Preface to the second edition

This book was written 15 years ago; literally half a lifetime ago, for me. In the intervening years, I have left home, moved to a different country, gone to graduate school, taught classes, written research papers, advised graduate students, married my wife, and had a son. Clearly, my perspective on life and on mathematics is different now than it was when I was 15. I have not been involved in problem-solving competitions for a very long time now, and if I were to write a book now on the subject it would be very different from the one you are reading here.

Mathematics is a multifaceted subject, and our experience and appreciation of it changes with time and experience. As a primary school student, I was drawn to mathematics by the abstract beauty of formal manipulation, and the remarkable ability to repeatedly use simple rules to achieve non-trivial answers. As a high-school student, competing in mathematics competitions, I enjoyed mathematics as a sport, taking cleverly designed mathematical puzzle problems (such as those in this book) and searching for the right 'trick' that would unlock each one. As an undergraduate, I was awed by my first glimpses of the rich, deep, and fascinating theories and structures which lie at the core of modern mathematics today. As a graduate student, I learnt the pride of having one's own research project, and the unique satisfaction that comes from creating an original argument that resolved a previously open question. Upon starting my career as a professional research mathematician, I began to see the intuition and motivation that lay behind the theories and problems of modern mathematics, and was delighted when realizing how even very complex and deep results are often at heart be guided by very simple, even common-sensical, principles. The 'Aha!' experience of grasping one of these principles, and suddenly seeing how it illuminates and informs a large body of mathematics, is a truly remarkable one. And there are yet more aspects of mathematics to discover; it is only recently for me that I have grasped enough fields of mathematics to begin to get a sense of the endeavour of modern mathematics as a unified subject, and how it connects to the sciences and other disciplines.

As I wrote this book before my professional mathematics career, many of these insights and experiences were not available to me, and so in many places the exposition has a certain innocence, or even naivety. I have been reluctant to tamper too much with this, as my younger self was almost certainly more attuned to the world of the high-school problem solver than I am now. However, I have made a number of organizational changes: formatting the text into LaTeX, arranging the material into what I believe is a more logical order, and editing those parts of the text which were inaccurate, badly worded, confusing, or unfocused. I have also added some more exercises. In some places, the text is a bit dated (Fermat's last theorem, for instance, has now been proved rigorously), and I now realize that several of the problems here could be handled more quickly and cleanly by more 'high-tech' mathematical tools; but the point of this text is not to present the slickest solution to a problem or to provide the most up-to-date survey of results, but rather to show how one approaches a mathematical problem for the first time, and how the painstaking, systematic experience of trying some ideas, eliminating others, and steadily manipulating the problem can lead, ultimately, to a satisfying solution.

I am greatly indebted to Tony Gardiner for encouraging and supporting the reprinting of this book, and to my parents for all their support over the years. I am also touched by all the friends and acquaintances I have met over the years who had read the first edition of the book. Last, but not least, I owe a special debt to my parents and the Flinders Medical Centre computer support unit for retrieving a 15-year old electronic copy of this book from our venerable Macintosh Plus computer!

Terence Tao
Department of Mathematics,
University of California, Los Angeles
December 2005

1 Strategies in problem solving

> The journey of a thousand miles begins with one step.
> Lao Tzu

Like and unlike the proverb above, the solution to a problem begins (and continues, and ends) with simple, logical steps. But as long as one steps in a firm, clear direction, with long strides and sharp vision, one would need far, far less than the millions of steps needed to journey a thousand miles. And mathematics, being abstract, has no physical constraints; one can always restart from scratch, try new avenues of attack, or backtrack at an instant's notice. One does not always have these luxuries in other forms of problem-solving (e.g. trying to go home if you are lost).

Of course, this does not necessarily make it easy; if it was easy, then this book would be substantially shorter. But it makes it possible.

There are several general strategies and perspectives to solve a problem correctly; (Polya 1957) is a classic reference for many of these. Some of these strategies are discussed below, together with a brief illustration of how each strategy can be used on the following problem:

PROBLEM 1.1. A triangle has its lengths in an arithmetic progression, with difference d. The area of the triangle is t. Find the lengths and angles of the triangle.

Understand the problem. What kind of problem is it? There are three main types of problems:

- 'Show that ...' or 'Evaluate ...' questions, in which a certain statement has to be proved true, or a certain expression has to be worked out;
- 'Find a ...' or 'Find all ...' questions, which requires one to find something (or everything) that satisfies certain requirements;
- 'Is there a ...' questions, which either require you to prove a statement or provide a counterexample (and thus is one of the previous two types of problem).

The type of problem is important because it determines the basic method of approach. 'Show that ...' or 'Evaluate ...' problems start with given data and the objective is to deduce some statement or find the value of an expression; this type of problem is generally easier than the other two types because there is a clearly visible objective, one that can be deliberately approached. 'Find a ...' questions are more hit-and-miss; generally one has to guess one answer that nearly works, and then tweak it a bit to make it more correct; or alternatively one can alter the requirements that the object-to-find must satisfy, so that they are easier to satisfy. 'Is there a ...' problems are typically the hardest, because one must first make a decision on whether an object exists or not, and provide a proof on one hand, or a counter-example on the other.

Of course, not all questions fall into these neat categories; but the general format of any question will still indicate the basic strategy to pursue when solving a problem. For example, if one tries to solve the problem 'find a hotel in this city to sleep in for the night', one should alter the requirements to, say 'find a vacant hotel within 5 kilometres with a room that costs less than 100$ a night' and then use pure elimination. This is a better strategy than proving that such a hotel does or does not exist, and is probably a better strategy than picking any handy hotel and trying to prove that one can sleep in it.

In Problem 1.1 question, we have an 'Evaluate ...' type of problem. We need to find several unknowns, given other variables. This suggests an algebraic solution rather than a geometric one, with a lot of equations connecting d, t, and the sides and angles of the triangle, and eventually solving for our unknowns.

Understand the data. What is given in the problem? Usually, a question talks about a number of objects which satisfy some special requirements. To understand the data, one needs to see how the objects and requirements react to each other. This is important in focusing attention on the proper techniques and notation to handle the problem. For example, in our sample question, our data are a triangle, the area of the triangle, and the fact that the sides are in an arithmetic progression with separation d. Because we have a triangle, and are considering the sides and area of it, we would need theorems relating sides, angles, and areas to tackle the question: the sine rule, cosine rule, and the area formulas, for example. Also, we are dealing with an arithmetic progression, so we would need some notation to account for that; for example, the side lengths could be a, $a + d$, and $a + 2d$.

Understand the objective. What do we want? One may need to find an object, prove a statement, determine the existence of an object with special properties, or whatever. Like the flip side of this strategy, 'understand the data', knowing the objective helps focus attention on the best weapons to use. Knowing the objective also helps in creating tactical goals which we know will bring us closer to solving the question. Our example question has the objective of 'find all the sides and angles of the triangle'. This means, as mentioned before, that we will need theorems and results concerning sides and angles. It also gives us the tactical goal of 'find equations involving the sides and angles of the triangle'.

Select good notation. Now that we have our data and objective, we must represent it in an efficient way, so that the data and objective are both represented as simply as possible. This usually involves the thoughts of the past two strategies. In our sample question, we are already thinking of equations involving d, t, and the sides and angles of the triangle. We need to express the sides and angles in terms of variables: one could choose the sides to be a, b, and c, while the angles could be denoted α, β, γ. But we can use the data to simplify the notation: we know that the sides are in arithmetic progression, so instead of a, b, and c, we can have a, $a + d$, and $a + 2d$ instead. But the notation can be even better if we make it more symmetrical, by making the side lengths $b - d$, b, and $b + d$. The only slight drawback to this notation is that b is forced to be larger than d. But on further thought, we see that this is actually not a restriction; in fact the knowledge that $b > d$ is an extra piece of data for us. We can also trim the notation more, by labelling the angles α, β, and $180° - \alpha - \beta$, but this is ugly and unsymmetrical—it is probably better to keep the old notation, but bearing in mind that $\alpha + \beta + \gamma = 180°$.

Write down what you know in the notation selected; draw a diagram. Putting everything down on paper helps in three ways:

(a) you have an easy reference later on;
(b) the paper is a good thing to stare at when you are stuck;
(c) the physical act of writing down of what you know can trigger new inspirations and connections.

Be careful, though, of writing superfluous material, and do not overload your paper with minutiae; one compromise is to highlight those facts which you think will be most useful, and put more questionable, redundant, or crazy ideas in another part of your scratch paper. Here are some equations and inequalities one can extract from our example question:

- (physical constraints) $\alpha, \beta, \gamma, t > 0$, and $b \geq d$; we can also assume $d \geq 0$ without loss of generality;
- (sum of angles in a triangle) $\alpha + \beta + \gamma = 180°$;
- (sine rule) $(b-d)/\sin\alpha = b/\sin\beta = (b+d)/\sin\gamma$;
- (cosine rule) $b^2 = (b-d)^2 + (b+d)^2 - 2(b-d)(b+d)\cos\beta$, etc.;
- (area formula) $t = (1/2)(b-d)b\sin\gamma = (1/2)(b-d)(b+d)\sin\beta = (1/2)b(b+d)\sin\alpha$;
- (Heron's formula) $t^2 = s(s-b+d)(s-b)(s-b-d)$, where $s = ((b-d)+b+(b+d))/2$ is the semiperimeter;
- (triangle inequality) $b + d \leq b + (b-d)$.

Many of these facts may prove to be useless or distracting. But we can use some judgement to separate the valuable facts from the unhelpful ones. The equalities are likely to be more useful than the inequalities, since our objective and data come in the form of equalities. And Heron's formula looks especially promising, because the semiperimeter simplifies to $s = 3b/2$. So we can highlight 'Heron's formula' as being likely to be useful.

We can of course also draw a picture. This is often quite helpful for geometry questions, though in this case the picture does not seem to add much:

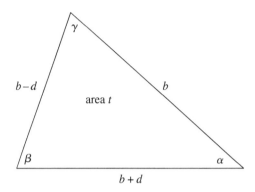

Modify the problem slightly. There are many ways to vary a problem into one which may be easier to deal with:

(a) Consider a special case of the problem, such as extreme or degenerate cases.
(b) Solve a simplified version of the problem.
(c) Formulate a conjecture which would imply the problem, and try to prove that first.
(d) Derive some consequence of the problem, and try to prove that first.
(e) Reformulate the problem (e.g. take the contrapositive, prove by contradiction, or try some substitution).
(f) Examine solutions of similar problems.
(g) Generalize the problem.

This is useful when you cannot even get started on a problem, because solving for a simpler related problem sometimes reveals the way to go on the main problem. Similarly, considering extreme cases and solving the problem with additional assumptions can also shed light on the general solution. But be warned that special cases are, by their nature, special, and some elegant technique could conceivably apply to them and yet have absolutely no utility in solving the general case. This tends to happen when the special case is *too* special. Start with modest assumptions first, because then you are sticking as closely as possible to the spirit of the problem.

In Problem 1.1, we can try a special case such as $d = 0$. In this case we need to find the side length of an equilateral triangle of area t. In this case, it is a standard matter to compute the answer, which is $b = 2t^{1/2}/3^{1/4}$. This indicates that the general answer should also involve square roots or fourth roots, but does not otherwise suggest how to go about the problem. Consideration of similar problems draws little as well, except one gets further evidence that a gung-ho algebraic attack is what is needed.

Modify the problem significantly. In this more aggressive type of strategy, we perform major modifications to a problem such as removing data, swapping the data with the objective, or negating the objective (e.g. trying to disprove a statement rather than prove it). Basically, we try to push the problem until it breaks, and then try to identify where the breakdown occurred; this identifies what the key components of the data are, as well as where the main difficulty will lie. These exercises can also help in getting an instinctive feel of what strategies are likely to work, and which ones are likely to fail.

In regard to our particular question, one could replace the triangle with a quadrilateral, circle, etc. Not much help there: the problem just gets more complicated. But on the other hand, one can see that one does not really need a triangle in the question, but just the dimensions of the triangle. We do not really need to know the position of the triangle. So here is further confirmation that we should concentrate on the sides and angles (i.e. $a, b, c, \alpha, \beta, \gamma$) and not on coordinate geometry or similar approaches.

We could omit some objectives; for example, instead of working out all the sides and angles we could work out just the sides, for example. But then one can notice that by the cosine and sine rules, the angles of the triangle will be determined anyway. So it is only neccesary to solve for the sides. But we know that the sides have lengths $b - d$, b, and $b + d$, so we only need to find what b is to finish the problem.

We can also omit some data, like the arithmetic difference d, but then we seem to have several possible solutions, and not enough data to solve the problem. Similarly, omitting the area t will not leave enough data to clinch a solution. (Sometimes one can *partially* omit data, for instance, by only specifying that the area is larger or smaller than some threshold t_0; but this is getting complicated. Stick with the simple options first.) Reversal of the problem (swapping data with objective) leads to some interesting ideas though. Suppose you had a triangle with arithmetic difference d, and you wanted to shrink it (or whatever) until the area becomes t. One could imagine our triangle shrinking and deforming, while preserving the arithmetic difference of the sides. Similarly, one could consider all triangles with a fixed area, and mold the triangle into one with the sides in the correct arithmetic progression. These ideas could work in the long run: but I will solve this question by another approach. Do not forget, though, that a question can be solved in more than one way, and no particular way can really be judged the absolute best.

Prove results about our question. Data is there to be used, so one should pick up the data and play with it. Can it produce more meaningful data? Also, proving small results could be beneficial later on, when trying to prove the main result or to find the answer. However small the result, do not forget it—it could have bearing later on. Besides, it gives you something to do if you are stuck.

In a 'Evaluate ...' problem like the triangle question, this tactic is not as useful. But one can try. For example, our tactical goal is to solve for b. This depends on the two parameters d and t. In other words, b is really a function: $b = b(d, t)$. (If this notation looks out of place in a geometry question, then that is only because geometry tends to ignore the functional dependence of objects. For example, Heron's formula gives an explicit form for the area A in terms of the sides a, b, and c: in other words, it expresses the function $A(a, b, c)$.) Now we can prove some mini-results about this function $b(d, t)$, such as $b(d, t) = b(-d, t)$ (because an arithmetic progression has an equivalent arithmetic progression with inverted arithmetic difference), or $b(kd, k^2 t) = kb(d, t)$ (this is done by dilating the triangle that satisfies $b = b(d, t)$ by k). We could even try differentiate b with respect to d or t. For this particular problem, these tactics allow us to perform some normalizations, for instance setting $t = 1$ or $d = 1$, and also provide a way to check the final answer. However, in this problem these tricks turn out to only give minor advantages and we will not use them here.

Simplify, exploit data, and reach tactical goals. Now we have set up notation and have a few equations, we should seriously look at attaining our tactical goals that we have established. In simple problems, there are usually standard ways of doing this. (For example, algebraic simplification is usually discussed thoroughly in high-school level textbooks.) Generally, this part is the longest and most difficult part of the problem: however, once can avoid getting lost if one remembers the relevant theorems, the data and how they can be used, and most importantly the objective. It is also a good idea to not apply any given technique or method blindly, but to think ahead and see where one could hope such a technique to take one; this can allow one to save enormous amounts of time by eliminating unprofitable directions of inquiry before sinking lots of effort into them, and conversely to give the most promising directions priority.

In Problem 1.1, we are already concentrating on Heron's formula. We can use this to attain our tactical goal of solving for b. After all, we have already noted that the sine and cosine rules can determine α, β, γ once b is known. As further evidence that this is going to be a step forward, note that Herons formula involves d and t—in essence, it uses all our data (we have already incorporated the fact about the sides being in arithmetic progression into our notation). Anyway, Herons formula in terms of d, t, b becomes

$$t^2 = \frac{3b}{2}\left(\frac{3b}{2} - b + d\right)\left(\frac{3b}{2} - b\right)\left(\frac{3b}{2} - b - d\right)$$

which we can simplify to

$$t^2 = \frac{3b^2(b-2d)(b+2d)}{16} = \frac{3b^2(b^2 - 4d^2)}{16}.$$

Now we have to solve for b. The right-hand side is a polynomial in b (treating d and t as constants), and in fact it is a quadratic in b^2. Now quadratics can be solved easily: if we put clear denominators and put everything on the left-hand side we get

$$3b^4 - 12d^2b^2 - 16t^2 = 0$$

so, using the quadratic formula,

$$b^2 = \frac{12d^2 \pm \sqrt{144d^4 + 196t^2}}{6} = 2d^2 \pm \sqrt{4d^2 + \frac{16}{3}t^2}.$$

Because b has to be positive, we get

$$b = \sqrt{2d^2 + \sqrt{4d^4 + \frac{16}{3}t^2}},$$

as a check, we can verify that when $d = 0$ this agrees with our previous computation of $b = 2t^{1/2}/3^{1/4}$. Once we ompute the sides $b - d, b, b + d$, the evaluation of the angles α, β, γ then follows from the cosine laws, and we are done!

2 Examples in number theory

> There is divinity in odd numbers, either in nativity, chance, or death.
> *William Shakespeare*, The Merry Wives of Windsor

Number theory may not neccesarily be divine, but it still has an aura of mystique about it. Unlike algebra, which has as its backbone the laws of manipulating equations, number theory seems to derive its results from a source unknown. Take, for example, *Lagrange's theorem* (first conjectured by Fermat) that every positive integer is a sum of four perfect squares (e.g. $30 = 4^2 + 3^2 + 2^2 + 1^2$). Algebraically, we are talking about an extremely simple equation: but because we are restricted to the integers, the rules of algebra fail. The result is infuriatingly innocent-looking and experimentation shows that it does seem to work, but offers no explanation why. Indeed, Lagrange's theorem cannot be easily proved by the elementary means covered in this book: an excursion into Gaussian integers or something similar is needed.

Other problems, though, are not as deep. Here are some simple examples, all involving a natural number n:

(a) n always has the same last digit as its fifth power n^5.
(b) n is a multiple of 9 if and only if the sum of its digits is a multiple of 9.
(c) (Wilson's theorem) $(n-1)! + 1$ is a multiple of n if and only if n is a prime number.
(d) If k is a positive odd number, then $1^k + 2^k + \cdots + n^k$ is divisible by $n+1$.
(e) There are exactly four numbers that are n digits long (allowing for padding by zeroes) and which are exactly the same last digits as their square. For instance, the four three-digit numbers with this property are 000, 001, 625, and 876.

These statements can all be proved by elementary number theory; all revolve around the basic idea of *modular arithmetic*, which gives you the power of algebra but limited to a finite number of integers. Incidentally, trying to solve the last assertion (e) can eventually lead to the notion of *p-adics*, which is sort of an infinite-dimensional form of modular arithmetic.

2 : Examples in number theory

Basic number theory is a pleasant backwater of mathematics. But the applications that stem from the basic concepts of integers and divisibility are amazingly diverse and powerful. The concept of divisibility leads naturally to that of *primes*, which moves into the detailed nature of factorization and then to one of the jewels of mathematics in the last part of the previous century: the prime number theorem, which can predict the number of primes less than a given number to a good degree of accuracy. Meanwhile, the concept of integer operations lends itself to modular arithmetic, which can be generalized from a subset of the integers to the algebra of finite groups, rings, and fields, and leads to algebraic number theory, when the concept of 'number' is expanded into irrational surds, elements of splitting fields, and complex numbers. Number theory is a fundamental cornerstone which supports a sizeable chunk of mathematics. And, of course, it is fun too.

Before we begin looking at problems, let us review some basic notation. A *natural number* is a positive integer (we will not consider 0 a natural number). The set of natural numbers will be denoted as $\mathbf{N}$. A *prime number* is a natural number with exactly two factors: itself and 1; we do not consider 1 to be prime. Two natural numbers m and n are *coprime* if their only common factor is 1.

The notation '$x = y \pmod{n}$', which we read as 'x equals y modulo n', means that x and y differ by a multiple of n, thus for instance $15 = 65 \pmod{10}$. The notation '$\pmod{n}$' signifies that we are working in a *modular arithmetic* where the *modulus* n has been identified with 0; thus for instance modular arithmetic (mod 10) is the arithmetic in which $10 = 0$. Thus, for instance, we have $65 = 15 + 10 + 10 + 10 + 10 + 10 = 15 + 0 + 0 + 0 + 0 + 0 = 15 \pmod{10}$. Modular arithmetic also differs from standard arithmetic in that inequalities do not exist, and that all numbers are integers. For example, $7/2 \neq 3.5 \pmod{5}$, but rather $7/2 = 12/2 = 6 \pmod{5}$ because $7 = 12 \pmod{5}$. It may seem strange to divide in this round-about way, but in fact one can find that there is no real contradiction, although some divisions are illegal, just as division-by-zero is illegal within the traditional field of real numbers. As a general rule, division is OK if the denominator is coprime with the modulus n.

2.1 Digits

We mentioned above that one can learn something about a number (in particular, whether it is divisible by 9) by summing all its digits. In higher mathematics, it turns out that this operation is not particularly important (it has proven far more effective to study numbers directly, rather than via their digit expansion), but it is quite popular in recreational mathematics and has even been given mystical connotations by some! Certainly, digit summing appears fairly often in mathematics competition problems, such as this one.

> PROBLEM 2.1 (Taylor 1989, p. 7). Show that among any 18 consecutive three-digit numbers there is at least one which is divisible by the sum of its digits.

This is a finite problem: there are only 900 or so three-digit numbers, so theoretically we could evaluate the problem manually. But let us see if we can save ourselves some work. First of all, the objective looks a little weird: we want the number to be divisible by the sum of digits. Let us first write down the objective as a mathematical formula, so that we can manipulate it more easily. A three-digit number can be written in the form abc_{10} where a, b, c are the digits; we are writing abc_{10} to avoid confusion with abc; note that $abc_{10} = 100a + 10b + c$, but $abc = a \times b \times c$. If we use the standard notation $a|b$ to denote the statement that a divides b, we now want to solve

$$(a + b + c) | abc_{10}, \tag{1}$$

where abc_{10} are the digits of one of the 18 given consecutive numbers. Can we reduce, simplify, or somehow make usable this equation? It is possible, but it is not simplifiable to anything halfway decent (e.g. a useful equation connecting a, b, and c directly). In fact (1) is a horrendous thing to manipulate, even after one substitutes $100a + 10b + c$ for abc_{10}. Take a look at the solutions abc_{10} of (1):

$$100, 102, 108, 110, 111, 112, 114, 117, 120, 126, \ldots, 990, 999.$$

They seem to be haphazard and random. However, they do seem to occur often enough so that any run of 18 consecutive numbers should have one. And what is the significance of the 18 anyway? Assuming it is not a red herring, (perhaps only 13 consecutive numbers are needed, but the 18 is there to throw you off the track) why have 18? It may occur to some that the sums of digits of a number are rather related to the number 9 (e.g. any number has the same remainder as its digit sum upon dividing by 9) and 18 is related to 9, so there could be a vague connection. Still, consecutive numbers and divisibility do not mix. It seems that we have to reformulate the question or propose a related one to have a hope of solving it.

Now that we are on the lookout for anything related to the number 9, we should notice that most numbers which actually do satisfy (1) are multiples of 9, or at least multiples of 3. In fact there are only three exceptions on the list above (100, 110, and 112), and practically all of the multiples of 9 satisfy (1). So instead of trying to prove

> For any 18 consecutive numbers, at least one solves (1).

directly, we could try something like

> For any 18 consecutive numbers, there is a multiple of 9 which solves (1).

This route seems to 'break the ice' between our data (18 consecutive numbers) and the objective (a number satisfying (1)) because 18 consecutive numbers always contain a multiple of 9 (in fact they contain two such multiples), and from numerical evidence, and the heuristic properties of the number 9, it seems that multiples of 9 satisfy (1). This 'stepping stone' approach is the best way to reconcile two unfriendly statements.

Now this particular stepping stone (considering multiples of 9) does work, but a bit of extra work is needed to cover all the cases. It is actually better to use multiples of 18:

> 18 consecutive numbers $\implies$ a multiple of 18 $\implies$ a solution to (1)

The reasons for this change are twofold:

- 18 consecutive numbers will always contain exactly one multiple of 18, but they would contain two multiples of 9. It seems neater, and more appropriate, to use multiples of 18 than to use multiples of 9. After all, if we use multiples of 9 to solve the problem, the question would only need 9 consecutive numbers instead of 18.
- It should be easier to prove (1) for multiples of 18 than for multiples of 9, since multiples of 18 are nothing more than a special case of multiples of 9. Indeed, it turns out that multiples of 9 do not always work (consider for instance 909), but multiples of 18 will, as we shall see.

Anyway, experimentation shows that multiples of 18 seem to work. But why? Take, for example, 216, which is a multiple of 18. The sum of digits is 9, and 9 divides 216 because 18 divides 216. To consider another example: 882 is a multiple of 18, and the sum of digits is 18. Hence 882 is obviously divisible by its digit sum. Messing around with a few more examples shows that the sum of digits of a multiple of 18 is always 9 or 18, which divides the original number almost by default. And with these guesses a proof soon follows:

PROOF. Within the 18 consecutive numbers, one must be a multiple of 18, say abc_{10}. Because abc_{10} is a multiple of 9 as well, $a+b+c$ must be a multiple of 9. (Remember the divisibility rule for 9? A number is divisible by 9 if and only if its digit sum is divisible by 9). Because $a+b+c$ ranges between 1 and 27, $a+b+c$ must be 9, 18, or 27. 27 only occurs when $abc = 999$, but that is not a multiple of 18. Hence $a+b+c$ is 9 or 18, and so $a+b+c|18$. But $18|abc_{10}$ by definition, so $a+b+c|abc_{10}$, as desired. □

Remember that with questions involving things like digits, a direct approach is not usually the answer. A cumbersome formula should be simplified into something more manageable. In this case, the phrase 'one number out of any 18 consecutive numbers must be' was replaced by 'any multiple of 18 must be' which was weaker, but simpler and more relevant to the question (which was related to divisibility). It turned out to be a good guess, though. And remember that with finite problems, the strategies are not like those in higher mathematics. For example, the formula

$$a+b+c|abc_{10}$$

was not treated like typical mathematics (e.g. application of modular arithmetic), but instead we placed bounds on $a+b+c$ (9, 18, or 27) due to the fact that all numbers had only three digits, leaving us with the much simpler

$$9|abc_{10}, \quad 18|abc_{10}, \quad \text{or } 27|abc_{10}$$

Indeed, we never even had to expand out abc_{10} algebraically as $100a + 10b + c$; while that may have seemed like the logical first step, it turns out that it is sort of a red herring and does not make the problem any clearer to solve.

A final remark: It turns out that 18 consecutive numbers are the least number needed to insure one of them satisfies (1). Seventeen numbers would not work; consider for instance the sequence from 559 to 575. (I used a computer for that, not some tricky mathematics.) Of course, one does not need to know this fact in order to solve the problem.

EXERCISE 2.1. In a parlour game, the 'magician' asks one of the participants to think of a three-digit number abc_{10}. Then the magician asks the participant to add the five numbers acb_{10}, bac_{10}, bca_{10}, cab_{10}, and cba_{10}, and reveal their sum. Suppose the sum was 3194. What was abc_{10}

originally? (Hint: Get a better expression for the sum of the five numbers, something more mathematical. Then use modular arithmetic to get some bounds on a, b, and c.)

PROBLEM 2.2 (Taylor 1989, p. 37). Is there a power of 2 such that its digits could be rearranged and made into another power of 2? (No zeroes are allowed in the leading digit: for example, 0032 is not allowed.)

This seems like an unsolvable combination: powers of 2, and digit rearranging. This is because

(a) digit rearrangement has so many possibilities;
(b) it is not easy to determine individual digits of a power of 2.

This probably means that something sneaky is needed.

The first sneaky thing to be done is to guess the answer. Circumstantial evidence (this problem is from a mathematics competition) suggests that this is not a trial-and-error question, and so the answer should probably be 'no'. (On the other hand, some exceptionally ingenious construction could pull off a clever rearrangement of digits—but such a construction is probably not easy to find. Guess the easy options first. If you are right, you have saved a lot of time by not pursuing the hard ways. If you are wrong, you were doomed to a long haul anyway. This does not mean that you should forget about a promising but hard way to solve the problem: but rather, to take a sensible look around before plunging into deep water.)

Like in Problem 2.1, the digits are really sort of a red herring. In Problem 2.1, we only wanted to know two things about the sum of the digits: first, a divisibility condition, and second a size restriction. We did not want to introduce all the complications of an exact equation. It will probably be much the same here: we have to simplify the problem by generalizing the digit-switching process. From a purely logical viewpoint, we are worse off because we have to prove more: but in terms of clarity and simplicity we are gaining ground. (Why burden yourself with data that cannot be used? It will just be a distraction.)

So, we now have to pick out the main properties of powers of 2 and digit-switching – hopefully, we will find properties of one that are incompatible with the other. Now let us tackle powers of 2 first; they are easier to handle. Here they are:

$$1, 2, 4, 8, 16, 32, 64, 128, 256, 512, 1024, 2048, 4096, 8192,$$
$$16384, 32768, 65536, \ldots$$

Well, there is not very much you can say about the digits here. The last digit of a power of 2 is obviously even (except for the number 1), but the other digits are quite random-looking. Suppose you took the number 4096, for instance. An odd digit, a few even digits, and even a 0 digit here. What is stopping it being rearranged into another power of 2? Could it be rearranged into $2^{4256} = 1523\ldots936$, for instance? 'Of course not!' you would say. Why? 'Because it's far too big!'. So, does size count? 'Yes—There would be thousands of digits in 2^{4256}, and only four digits in 4096.' Aha—so rearranging digits cannot change the total number of digits. (Write down any facts which could be of use to your problem, even if they are simple—do not assume that 'obvious' facts will always spring to mind when needed. Even shallowly dug gold has to be searched—and held on to.)

Well, with this iota of information, can we proceed with our generalizing plan? Our generalized question is now

> Is there a power of 2 such that there is another power of 2 with the same number of digits as the first power of 2?

Unfortunately, the answer to this question is quickly seen to be 'yes'; 2048 and 4096, for example. We were too general. (Note that a 'yes' answer to this question does not necessarily yield a 'yes' answer to the original problem.) Again, look to Problem 2.1. Merely knowing 'the sum of digits of a multiple of 18 has to be a multiple of 9' is not sufficient to solve the problem: we also needed the fact that 'the sum of digits of three-digit number is at most 27'. In short, we have not found enough facts about our problem to solve it. Yet, we are still partially successful, because we have restricted the possibilities of digit rearranging. Take the number 4096 again. This can only be rearranged into another four-digit number. And how many four-digit powers of 2 are there? Only four—1024, 2048, 4096, and 8192. This is because the powers of 2 keep doubling: they can not stay in the same 'tax bracket' for too long. In fact, one can soon see that at most four powers of 2 can have the same number of digits. (The fifth consecutive power of 2 would be 16 times that of the first, and hence would have to have more digits than the first power of 2). So what this means is that for each power of 2, there are at most three other powers of 2 that could possibly be digit-rearrangements of the original power of 2. A partial victory: only three or fewer suspects left to eliminate for each power of 2, instead of the infinite number we had before. Perhaps with a bit of extra work we can eliminate those suspects as well.

We have said that when we switch the digits, the number you end up with has the same number of digits as the original. But the reverse is far from true, and this lone property of digit-switching will not solve the problem on its own. This means that we have generalized too far and pushed our luck too much. Let us reel ourselves in again. Something else could be preserved when we switch digits. Let us take a look at some examples—let us take 4096 again, since we have already got some experience with this number. The digit-switching possibilities are

$$4069, 4096, 4609, 4690, 4906, 4960, 6049, 6094, 6409,$$
$$6490, 6904, 6940, 9046, 9064, 9406, 9460, 9604, 9640.$$

What do they have in common? They have the same set of digits. That is all very well and good, but the 'set of digits' is not a very useful mathematical object (not many theorems and tools use this concept). However, the *sum of digits* is a more conventional weapon. And, well, if two numbers have the same set of digits, then they have to have the same digit-sum. So we have another iota of information: digit-switching preserves the digit-sum. Combining this with our previous iota we have a new replacement question:

> Is there a power of 2 such that there is another power of 2 with the same number of digits *and* the same digit-sum as the first power of 2?

Again, if this question is true, the original question is true. Now this question is a bit easier to cope with than the original, because 'number of digits' and 'digit-sums' are standard number-theory stuff.

With this new concept in mind, let us look at the digit-sums of the powers of 2, seeing as our new question involves them. Well, we have

Power of 2	Digit-sum	Power of 2	Digit-sum	Power of 2	Digit-sum
1	1	256	13	65,536	25
2	2	512	8	131,072	14
4	4	1,024	7	262,144	19
8	8	2,048	14	524,288	29
16	7	4,096	19	1,048,576	31
32	5	8,192	20		
64	10	16,384	22		
128	11	32,768	26		

From this we note that

- The digit-sums tend to be quite small. For instance, the digit-sum of 2^{17} is a mere 14. This is actually a small bit of bad luck, because small numbers are more likely to match than are big numbers. (If 10 people each randomly pick one two-digit number, there is a sizeable (9.5%) chance of a match, but if they each pick 10-digit numbers, then there is only a one in a million chance of a match: something about as lousy as the chances of winning the lottery.) But the smallness of the numbers also aids in picking out patterns, so perhaps it is not all bad news.

- Some digit-sums match: for example, 16 and 1024. But it seems that the digit-sums slowly climbs higher anyway: you would expect that a 100-digit power of 2 would have a higher digit-sum than a 10-digit one. But also remember that we are confining ourselves to powers of 2 with the same number of digits, so this idea will not be of much help.

The upshot of these observations is this: digit-sums have an easily appreciable macroscopic structure (slowly increasing with n; in fact it is highly probable (though not proven!) that the digit-sum of 2^n is approximately $(4.5 \log_{10} 2)n \approx 1.355n$ for large n) but a lousy microscopic structure. The digits just fluctuate too much. We mentioned earlier that 'set of digits' was unwieldy: now it seems that 'digit-sum' is not so flash either. Is there another reduction of the problem that will leave us with something we can really work with?

Hmm. We mentioned earlier that 'digit-sum' was a 'conventional weapon' in mathematics. Take a look at the preceding question for instance. But the only real way digit-sums can be successfully 'mainstreamed' is by considering the digit-sum modulo 9. One may recall that a number is equal to its digit-sum modulo 9; for example,

$$
\begin{aligned}
3297 &= 3 \times 10^3 + 2 \times 10^2 + 9 \times 10^1 + 7 \times 10^0 \pmod 9 \\
&= 3 \times 1^3 + 2 \times 1^2 + 9 \times 1^1 + 7 \times 1^1 \pmod 9 \\
&= 3 + 2 + 9 + 7 \pmod 9
\end{aligned}
$$

because 10 is equal to 1 (mod 9).

So now our new modified question is as follows:

> Is there a power of 2 such that there is another power of 2 with the same number of digits and the same digit-sum *modulo* 9 as the first power of 2?

Now we can use the fact that a number is equal to its digit-sum modulo 9 to rephrase this question again:

> Is there a power of 2 such that there is another power of 2 with the same number of digits and the same remainder (mod 9) as the first power of 2?

Note that the pesky notions of 'rearranging digits', 'set of digits', and 'sum of digits' have been completely eliminated, which looks promising.

2 : Examples in number theory

Now let us modify the above table of digit-sums of powers of 2 and see what we get.

Power of 2	(mod 9) Remainder	Power of 2	(mod 9) Remainder	Power of 2	(mod 9) Remainder
1	1	256	4	65,536	7
2	2	512	8	131,072	5
4	4	1,024	7	262,144	1
8	8	2,048	5	524,288	2
16	7	4,096	1	1,048,576	4
32	5	8,192	2		
64	1	16,384	4		
128	2	32,768	8		

What we have to prove is that no two powers of 2 have the same remainder (mod 9) and the same number of digits. Well, looking at the table, there are several powers of 2 with the same remainder: 1, 64, 4096, and 262144 for example. But none of these four have the same number of digits. Indeed, powers of 2 with the same remainder (mod 9) seem to be so separated that there is no hope of them having the same number of digits. In fact, the powers of 2 with the same remainder seem to be quite regularly spaced ... and one can quickly see that the remainders (mod 9) repeat themselves every six steps. This conjecture can be easily proved by modular arithmetic:

$$2^{n+6} = 2^n 2^6 = 2^n \times 64 = 2^n \pmod{9} \text{ because } 64 = 1 \pmod{9}.$$

This result means that the remainders of the powers of 2 will repeat themselves endlessly, like a repeating decimal: $1, 2, 4, 8, 7, 5, 1, 2, 4, 8, 7, 5, 1, 2, 4, 8, 7, 5, \ldots$. This in turn means that two powers of 2 with the same digit-sum (mod 9) must be at least six steps apart. But then the powers of 2 cannot possibly have the same number of digits, because one would be 64 times bigger than the other, at least. So this means that there are no powers of 2 with the same number of digits and the same digit-sum (mod 9). We have now proved our modified question, so we can work backwards until we reach our original question, and write out the full answer:

PROOF. Suppose two powers of 2 are related by digit-switching. This means that they have the same number of digits, and also have the same digit-sum (mod 9). But the digit-sums (mod 9) are periodic with a period of 6, with no repetitions within any given period, so the two powers are at least six steps apart. But then it is impossible for them to have the same number of digits, a contradiction. □

This problem was simplified repeatedly until the more unusable and unfriendly parts of the problem were exchanged with more natural, flexible, and co-operative notions. This simplification can be a bit of a hit-and-miss affair; there is always the danger of oversimplification, or mis-simplification (simplifying into a red herring). But in this question, almost anything was better than playing around with digit-switching, so simplification could not do much more harm. There is a chance that maneuvering and simplifying may land you into a wild goose chase, but if you are really stuck anyway, anything is worth a try.

2.2 Diophantine equations

A *Diophantine equation* is an algebraic equation (the classic one is $a^2 + b^2 = c^2$) with the constraint that all variables are integers. The usual objective is to find all solutions to the equation. Generally, there is more than one solution, even if everything is integral. These equations can be solved algebraically, but one also can use the number-theoretical methods of integer division, modular arithmetic, and integral factorization. Here is one:

> PROBLEM 2.3 (Australian Mathematics Competition 1987, p. 15).
> Find all integers n such that the equation $1/a + 1/b = n/(a+b)$ is satisfied for some non-zero integer values of a and b (with $a + b \neq 0$).

This seems like a standard Diophantine equation, so we would probably begin by multiplying out the denominators, to get

$$(a+b)/ab = n/(a+b)$$

and then

$$(a+b)^2 = nab. \qquad (2)$$

Now what? We could eliminate the n, and say that

$$ab \mid (a+b)^2$$

(using the divisibility symbol | that we used in Problem 2.1) or try to concentrate on the fact that nab is a square. These techniques are good, but they do not seem to work on this problem. The relationships of the left and right sides of (2) are not strong enough. One side is a square, the other is a product.

2: Examples in number theory

One thing to keep in mind when problem-solving is to be prepared to abandon temporarily one interesting—but fruitless—approach and try a more promising one. One could try algebra to attack the problem, then re-apply number theory later if algebra failed to work. Expanding (2) and collecting terms we can get

$$a^2 + (2-n)ab + b^2 = 0,$$

and if one is brave enough to use the quadratic formula we get

$$a = \frac{b}{2}\left[(n-2) \pm \sqrt{(n-2)^2 - 4}\right].$$

This looks very messy, but actually we can turn this messiness to our advantage. We know that a, b, and n are integers, but there is a square root in the formula. Now this can only work if the term inside the square root, $(n-2)^2 - 4$, is a perfect square. But this means that 4 less than a square is a square. This is very restrictive. Because the gaps between the squares get higher than 4 after the first few squares, we only need to test low numbers of n. It turns out that $(n-2)^2$ has to be 4, and hence n is either 0 or 4. Now we can work each case separately, finding either an example of each or a proof that no such example exists.

Case 1: $n = 0$. Feeding this back into, say, (2) we get $(a+b)^2 = 0$, and thus $a + b = 0$. But this is impossible as in our original equation we now have a 0/0, which is illegal. Hence n cannot be 0.

Case 2: $n = 4$. Again, (2) gives us $(a+b)^2 = 4ab$, which upon collecting terms gives $a^2 - 2ab + b^2 = 0$. Factorizing this we get $(a-b)^2 = 0$, so a must equal b. This is not a contradiction, but an example: $a = b$, $n = 4$, works when put into the original equation (2).

So our answer was $n = 4$, but it was obtained by the rather inelegant method of the quadratic formula. Using it is usually clumsy, but as it introduces a square root term, which implies that the term inside the square root must be a perfect square, it occasionally comes in useful.

Diophantine problems can get extremely difficult when one of the variables appears in the exponent; the most notorious of these is *Fermat's last theorem*, which asserts that there are no natural number solutions to $a^n + b^n = c^n$ with $n > 2$. Fortunately, there are other problems involving exponents which are easier to handle.

PROBLEM 2.4 (Taylor 1989, p.7). Find all solutions of $2^n + 7 = x^2$ where n and x are integers.

This kind of question really needs trial and error to find the right track. With Diophantine equations, the most elementary methods are modular arithmetic and factorization. Modular arithmetic transfers the entire equation to a suitable modulus, sometimes constant (e.g. (mod 7), or (mod 16)) or sometimes variable (e.g. (mod pq)). Factorization alters the problem into the form (factor) × (factor) = (something nice), where the right-hand side could be a constant (the best possible result), a prime, a square, or something else that has a limited choice of factors. For example, in Problem 2.3, both methods were considered early on, but discarded in favour of an algebraic approach, which is actually a factorization technique in disguise (remember we eventually got $(n-2)^2 - 4 =$ (square)?).

Now it is best to try elementary techniques first, as it may save a lot of dashing about in circles later. One may have abandoned these methods and tried to analyse the approximate equation

$$x = \sqrt{2^n + 7} \approx 2^{n/2}$$

which can get into some serious number theory involving topics such as continued fractions, Pell's equation, and recursion relations. It can be done; but we will look for the elegant (i.e. lazy) way out.

Obtaining a useful factorization is next to impossible, except when n is even. Then we get a difference of two squares (a vital factorization in Diophantine equations) like so:

$$7 = x^2 - 2^n = (x - 2^m)(x + 2^m),$$

where $m = n/2$. Then we can say that $x - 2^m$ and $x + 2^m$, being factors of 7, must be $-7, -1, 1,$ or 7; and further breakup into cases soon shows that there are no solutions (if we assume n is even). But that is about as much as the factorization method can tell us; it does not tell us where the actual solutions are and how many of them there are. (Although we do now know that n must be odd.)

The modular arithmetic approach is next. The strategy is to use the modulus to get rid of one or more of the terms. For example, we could write the equation modulo x, to obtain

$$2^n + 7 \equiv 0 \pmod{x},$$

or maybe modulo 7, to get

$$2^n \equiv x^2 \pmod{7}.$$

Unfortunately, these methods do not work well at all. But before we give up, there is one more modulus to try. We tried eliminating the '7' and the 'x^2' terms; can we eliminate the 2^n term instead? Yes, by choosing, say, mod 2. Then we get

$$0 + 7 = x^2 \text{ (mod 2)}$$

when $n > 0$, and

$$1 + 7 = x^2 \text{ (mod 2)}$$

when $n = 0$. This is not too bad as we have almost eliminated the role of n completely. But it still does not work, as the x^2 term on the right-hand side could be 0 or 1, so we have not really excluded any possibilities. To restrict the values of x^2, we have to choose a different modulus. With this line of thought—to restrict the values on the right-hand-side—one now thinks to try modulus 4 instead of 2:

$$2^n + 7 = x^2 \text{ (mod 4)}.$$

In other words, we have

$$0 + 3 = x^2 \text{ (mod 4)} \quad \text{when } n > 1, \tag{3}$$
$$2 + 3 = x^2 \text{ (mod 4)} \quad \text{when } n = 1, \tag{4}$$
$$1 + 3 = x^2 \text{ (mod 4)} \quad \text{when } n = 0. \tag{5}$$

Because x^2 must be 0 (mod 4) or 1 (mod 4), possibility (3) is eliminated. This means n can only be 0 or 1. A quick check then shows that only $n = 1$ can work, and x must be $+3$ or -3.

The main idea, when solving Diophantine equations of the form 'find all solutions', is to eliminate all but a finite number of possibilities. This is another reason why the (mod 7) and (mod x) would not work; for if they did, they would have eliminated all the cases, unlike the (mod 4) approach, which eliminated all but a handful.

EXERCISE 2.2. Find the largest positive integer n such that $n^3 + 100$ is divisible by $n + 10$. (Hint: use (mod $n + 10$). Get rid of the n by using the fact that $n = -10$ (mod $n + 10$).)

2.3 Sums of powers

> PROBLEM 2.5 (Hajós et al. 1963, p. 74). Prove that for any non-negative integer n, the number $1^n + 2^n + 3^n + 4^n$ is divisible by 5 if and only if n is not divisible by 4.

This problem looks a bit daunting at first: equations like the above may remind one of Fermat's last theorem, which is notorious for its insolvability. But our question is much milder. We wish to show that a certain number is (or is not) divisible by 5. Unless a direct factorization is evident, we will have to use the modulus approach. (That is, show that $1^n + 2^n + 3^n + 4^n = 0 \pmod 5$ for n not divisible by 4, and $1^n + 2^n + 3^n + 4^n \neq 0 \pmod 5$ otherwise.)

Because we are using such small numbers, we can evaluate some of the values of $1^n + 2^n + 3^n + 4^n \pmod 5$ manually. The best way to do this is to work out $1^n \pmod 5$, $2^n \pmod 5$, $3^n \pmod 5$, and $4^n \pmod 5$ individually before adding:

$\pmod 5$

n	1^n	2^n	3^n	4^n	$1^n + 2^n + 3^n + 4^n$
0	1	1	1	1	4
1	1	2	3	4	0
2	1	4	4	1	0
3	1	3	2	4	0
4	1	1	1	1	4
5	1	2	3	4	0
6	1	4	4	1	0
7	1	3	2	4	0
8	1	1	1	1	4

Now it is obvious that some periodicity is evident. In fact 1^n, 2^n, 3^n, and 4^n are all periodic with period 4. To prove this conjecture, we can just fiddle with the definition of periodicity.

Take 3^n, for example. Saying that this is periodic with period 4 just means that

$$3^{n+4} = 3^n \pmod 5.$$

But this is easy to prove, as

$$3^{n+4} = 3^n \times 81 = 3^n \pmod 5$$

because $81 = 1 \pmod 5$.

Similarly we can prove 1^n, 2^n, and 4^n are periodic with period 4. This means that $1^n + 2^n + 3^n + 4^n$ is periodic with period 4. This in turn implies that we only need to prove our question for $n = 0, 1, 2, 3$, because periodicity will take care of all the other cases of n. But we have already shown the question to be true in these cases (see the above table). So we are done. (By the way, there is a more elementary method available if we assume that n is odd: simply pair up and cancel terms.)

Whenever trying to prove equations involving a parameter (in this case n), periodicity is always handy, as one no longer needs to check all values of the parameter to verify the equation. Checking one period (e.g. $n = 0$, 1, 2, and 3) will be sufficient.

> EXERCISE 2.3. Show that the equation $x^4 + 131 = 3y^4$ has no solutions if x and y are integers.

Now we turn to a trickier problem concerning sums of powers.

> PROBLEM 2.6 (Shklarsky et al. 1962, p. 14). (**) Let k, n be natural numbers with k odd. Prove that the sum $1^k + 2^k + \cdots + n^k$ is divisible by $1 + 2 + \cdots + n$.

This question, by the way, is a standard exercise in Bernoulli polynomials (or some astute applications of the Remainder Theorem), an interesting portion of mathematics that has many applications. But without the sledge-hammer of Bernoulli polynomials (or the Riemann ζ function) we will just have to use plain old number theory.

First of all, we know that $1 + 2 + \cdots + n$ can also be written in the form $n(n+1)/2$. Which form shall we use? The former is more aesthetic, but a bit useless in a divisibility question. (It is always easier if the divisor is expressed as a product, rather than a sum.) It might have been useful if there was some nice factorization of $1^k + 2^k + \cdots + n^k$ which involved $1 + 2 + \cdots + n$, but there is not (at least, not an obvious one). If there was some way to relate divisibility by $1 + 2 + \cdots + n$ to divisibility by $1 + 2 + \cdots + (n+1)$ then induction might be a way to go, but that does not seem likely either. So we will try the $n(n+1)/2$ formulation instead.

So, using modular arithmetic (which is the most flexible way to prove that one number divides another), our objective is to show that

$$1^k + 2^k + \cdots + n^k \equiv 0 \pmod{n(n+1)/2}.$$

Let us ignore for the moment the '2' in the $n(n+1)/2$. Then we are trying to prove something of the form

$$\text{(factor 1)} \times \text{(factor 2)} | \text{(expression)}.$$

If the two factors are coprime, then our objective is equivalent to proving both of

$$\text{(factor 1)} | \text{(expression)} \text{ and } \text{(factor 2)} | \text{(expression)}$$

separately. This should be simpler to prove: it is easier to prove divisibility if the divisors are smaller. But there is an annoying '2' in the way. To deal with that we will just break up into cases, depending on whether n is even or odd.* The cases are quite similar and I will only do the case when n is even. In this case we can write $n = 2m$ (so as to avoid staring at messy '$n/2$' terms in the following equations—little housekeeping things like this help a solution run smoothly). Replacing all the ns by $2m$s, we have to prove

$$1^k + 2^k + \cdots + (2m)^k = 0 \ (\text{mod } m(2m+1)),$$

but since m and $2m+1$ are coprime, this is equivalent to proving

$$1^k + 2^k + \cdots + (2m)^k = 0 \ (\text{mod } 2m+1)$$

and

$$1^k + 2^k + \cdots + (2m)^k = 0 \ (\text{mod } m).$$

Let us tackle the (mod $2m+1$) part first. It is quite similar to Problem 2.5 but is a bit easier, because we know that k is odd. Using the modulus $2m+1$, $2m$ is equivalent to -1, $2m-1$ is equivalent to -2, and so on, so our expression $1^k + 2^k + \cdots + (2m)^k$ becomes

$$1^k + 2^k + \cdots + (m)^k + (-m)^k + \cdots + (-2)^k + (-1)^k \ (\text{mod } 2m+1).$$

We have done this so that we can do some nice cancelling. k is odd, so $(-1)^k$ is equal to -1. Therefore $(-a)^k = -a^k$. The upshot of this is that the above sum can be pairwise cancelled: 2^k and $(-2)^k$ will cancel, 3^k and $(-3)^k$ will cancel, etc., leaving 0 (mod $2m+1$), as desired.

---

* Another way is to multiply both sides by 2, so that we now want to prove $2(1^k + 2^k + \cdots + n^k) = 0 \ (\text{mod } n(n+1))$. This ends up being more or less equivalent to the approach given below.

Now we have to do the (mod m) part: that is, we have to show

$$1^k + 2^k + 3^k + \cdots + (m-1)^k + (m)^k + (m+1)^k + \cdots$$
$$+ (2m-1)^k + (2m)^k \equiv 0 \pmod{m}.$$

But we are working modulo m, so some of the above terms can be simplified. m and $2m$ are both equivalent to 0 (mod m), and $m+1$ is equivalent to 1, $m+2$ is equivalent to 2, and so on. So the above summation simplifies to

$$1^k + 2^k + 3^k + \cdots + (m-1)^k + 0^k + 1^k + \cdots + (m-1)^k + 0 \pmod{m}.$$

But several terms appear twice, so recombining (and ditching the 0s) we get

$$2(1^k + 2^k + 3^k + \cdots + (m-1)^k) \pmod{m}.$$

Now we can almost do the same thing as for the (mod $2m+1$) case, except there is a small hitch when m is even. If m is odd, we can reformulate the above expression as

$$2(1^k + 2^k + 3^k + \cdots + ((m-1)/2)^k + (-(m-1)/2)^k + \cdots$$
$$+ (-2)^k + (-1)^k) \pmod{m}.$$

and do the same procedure of cancellation as before. But if m is even (so $m = 2p$, say) there is a middle term, p^k, which does not cancel with anything. In other words, in this case the expression does not collapse to 0 immediately, but instead cancels to

$$2p^k \pmod{2p}.$$

But this, of course, is equal to 0. Regardless of whether m is odd or even, we have proved that $1^k + 2^k + 3^k + \cdots + n^k$ is divisible by $n(n+1)/2$ if n is even.

> EXERCISE 2.4. Complete the proof of the above problem by working out what happens when n is odd.

Now let us turn to a special type of 'sums of powers' problem, namely sums of reciprocals.

> **PROBLEM 2.7** (Shklarsky *et al.* 1962, p. 17). Let p be a prime number greater than 3. Show that the numerator of the (reduced) fraction
>
> $$1/1 + 1/2 + 1/3 + \cdots + 1/(p-1)$$
>
> is divisible by p^2. For example, when p is 5, the fraction is $1/1 + 1/2 + 1/3 + 1/4 = 25/12$, and the numerator is obviously divisible by 5^2.

This question is a 'Prove that' question, not a 'Find a' or 'Show there exists' question, so it should not be completely impossible. However, we have to prove something about a numerator of a reduced fraction—not something easily dealt with! This numerator will need to be transformed into something more standard, like an algebraic expression, so that we can manipulate it better. Also, the question does not just need divisibility by a prime, it needs divisibility by the square of a prime. This is significantly harder. We would like to somehow reduce the problem to mere prime divisibility to make the problem more solvable.

So by looking at the shape of the question, we have the following objectives to keep in mind:

(a) Express the numerator as a mathematical expression, so that we can manipulate it.
(b) Aim to reduce the problem from a p^2-divisiblity problem to something simpler, perhaps a p-divisibility problem.

Let us tackle (a) first. First of all, we can get a numerator easily, but not the reduced numerator necessarily. By adding up the fractions under a common denominator we get

$$\frac{(2 \times 3 \times \cdots \times (p-1)) + 1 \times 3 \times \cdots \times (p-1) + \cdots + 1 \times 2 \times 3 \times \cdots \times (p-2))}{(p-1)!}.$$

Now suppose that we can manage to prove that this numerator is divisible by p^2. How does this help us prove that the reduced numerator is also divisible by p^2? Well, what is the reduced numerator? It is the original numerator after some cancellation with the denominator. Can cancelling destroy the property of p^2-divisibility? Yes, if a multiple of p is cancelled. But multiples of p cannot be cancelled, because the denominator is coprime to p (p is prime, and $(p-1)!$ can be expressed as a product of numbers less than p). Aha! This means that we only need to prove that the ugly-looking numerator above is divisible by p^2. This is better than the other numerator because now we have an equation to solve:

$$2 \times 3 \times \cdots \times (p-1) + 1 \times 3 \times \cdots \times (p-1) + \cdots$$
$$+ 1 \times 2 \times 3 \times \cdots \times (p-2) = 0 \pmod{p^2}.$$

(Again, we have switched over to modular arithmetic, which is usually the best way to show that one number divides another. However, if the question involves more than one divisibility, for example, something involving all divisors of a certain number, other techniques are sometimes better.)

Although we have got an equation now, it is a mess. Our next task is to simplify it. What we have now on the left-hand side is an indefinite sum of indefinite products. (Indefinite just means that there are 'dot dot dots' in the expression.) However, we can represent the infinite products more neatly. Each infinite product is basically the numbers from 1 to $p-1$ multiplied together, except for one number, say i, which is between 1 and $p-1$. This can be expressed more compactly as $(p-1)!/i$; it is legitimate to divide by i modulo p^2 because i is coprime to p^2. So now our objective is to prove

$$\frac{(p-1)!}{1} + \frac{(p-1)!}{2} + \frac{(p-1)!}{3} + \cdots + \frac{(p-1)!}{p-1} = 0 \pmod{p^2}.$$

We factorize this to get

$$(p-1)!\left[\frac{1}{1} + \frac{1}{2} + \frac{1}{3} + \cdots + \frac{1}{p-1}\right] = 0 \pmod{p^2}. \quad (6)$$

(Remember that we are dealing with modular arithmetic, so that a number like 1/2 will be equivalent to an integer. For example, $1/2 = 6/2 = 3 \pmod{5}$.)

Now look at what we have: something of the form

$$(\text{factor}) \times (\text{factor}) = 0 \pmod{p^2}.$$

If it were not for the modular arithmetic, then we could quickly say that one of the factors is 0. With modular arithmetic, we can say nearly the same thing, but we have to be careful. Luckily, the first factor, $(p-1)!$, is coprime to p^2 (because $(p-1)!$ is coprime to p) so we can divide it out. The upshot of this is that (6) is equivalent to

$$\frac{1}{1} + \frac{1}{2} + \frac{1}{3} + \cdots + \frac{1}{p-1} = 0 \pmod{p^2}.$$

(Note that this looks very similar to our original question, the only difference being that we are considering the entire fraction, not just the numerator of it. But one cannot just jump from one form to another without care. The above complications were necessary.)

Now we have reduced the question to proving a rather benign-looking modular arithmetic equation. But where to go on from here? Perhaps an example will help. Let us take the same example as the one given in the question: namely, $p = 5$. We have

$$\frac{1}{1} + \frac{1}{2} + \frac{1}{3} + \frac{1}{4} = 1 + 13 + 17 + 19 \ (\text{mod } 25)$$
$$= 0 \ (\text{mod } 25)$$

as desired. But why does this work? The numbers 1, 13, 17, and 19 seem to be random, but 'magically' add up to the right amount. Perhaps it is a fluke. Let us try $p = 7$.

$$\frac{1}{1} + \frac{1}{2} + \frac{1}{3} + \frac{1}{4} + \frac{1}{5} + \frac{1}{6} = 1 + 25 + 33 + 37 + 10 + 41 \ (\text{mod } 49)$$
$$= 0 \ (\text{mod } 49).$$

This has the same 'flukiness' about it. How does this work? It is not clear how everything manages to cancel out modulo p^2. Perhaps, keeping objective (b) in mind, we can prove it (mod p) first, that is, let us first prove

$$\frac{1}{1} + \frac{1}{2} + \frac{1}{3} + \cdots + \frac{1}{p-1} = 0 \ (\text{mod } p). \tag{7}$$

If nothing else, it will give us something to do. (Besides, if we can not solve this (mod p) problem, there is no way that we will be able to solve the (mod p^2) problem.)

It turns out that the simpler problem (7) is much easier to work out. For example, when p is 5, we have

$$\frac{1}{1} + \frac{1}{2} + \frac{1}{3} + \frac{1}{4} = 1 + 3 + 2 + 4 \ (\text{mod } 5)$$
$$= 0 \ (\text{mod } 5),$$

while when p is 7 we have

$$\frac{1}{1} + \frac{1}{2} + \frac{1}{3} + \frac{1}{4} + \frac{1}{5} + \frac{1}{6} \ (\text{mod } 7) = 1 + 4 + 5 + 2 + 3 + 6 \ (\text{mod } 7)$$
$$= 1 + 2 + 3 + 4 + 5 + 6 \ (\text{mod } 7)$$
$$= 0 \ (\text{mod } 7).$$

Now we have a pattern emerging: the reciprocals $1/1, 1/2, \ldots, 1/(p-1)$ (mod p) seem to cover all the residues $1, 2, \ldots, (p-1)$ (mod p) exactly once. For example, in the above equation with $p = 7$, the numbers $1 + 4 + 5 + 2 + 3 + 6$ rearrange to form $1 + 2 + 3 + 4 + 5 + 6$, which is 0. To check a lengthier example, mod 11 yields

$$\frac{1}{1} + \frac{1}{2} + \cdots + \frac{1}{11} = 1 + 6 + 4 + 3 + 9 + 2 + 8 + 7 + 5 + 10 \text{ (mod 11)}$$
$$= 1 + 2 + 3 + 4 + 5 + 6 + 7 + 8 + 9 + 10 \text{ (mod 11)}$$
$$= 0.$$

This tactic, showing that the reciprocal numbers can be rearranged in this orderly fashion, works neatly for (mod p), but it does not generalize easily to (mod p^2). Instead of floundering around trying to fit a square block into a round hole (although it can be done if you push hard enough), it is better to find a block that is more round. So what we have to do now is find another proof of the fact that $\frac{1}{1} + \frac{1}{2} + \frac{1}{3} + \cdots + (1/p - 1) = 0$ (mod p); one that generalizes, at least partially, to the (mod p^2) case.

Now it is time to use experience with these sorts of problems. For example, if we are fresh from solving Problem 2.6, we know that symmetry, or anti-symmetry can be exploited, especially in modular arithmetic. In the problem of proving (7) we can make the sum more anti-symmetric by replacing $p - 1$ with -1, $p - 2$ with -2, and so forth, to get

$$\frac{1}{1} + \frac{1}{2} + \frac{1}{3} + \cdots + \frac{1}{p-1} = \frac{1}{1} + \frac{1}{2} + \frac{1}{3} + \cdots + \frac{1}{-3} + \frac{1}{-2} + \frac{1}{-1} \text{ (mod } p\text{)}.$$

And now we can pair off and cancel easily (there is no 'middle term' that does not pair off, as p is an odd prime). Can we do the same in (mod p^2)?

The answer is 'sort of'. When we solved the problem (mod p), we paired off $1/1$ and $1/(p-1)$, $1/2$, and $1/(p-2)$, and so forth. When we try the same pairing in (mod p^2), what we get now is this:

$$\frac{1}{1} + \frac{1}{2} + \cdots + \frac{1}{p-1}$$
$$= \left(\frac{1}{1} + \frac{1}{p-1}\right) + \left(\frac{1}{2} + \frac{1}{p-2}\right) + \cdots + \left(\frac{1}{(p-1)/2} + \frac{1}{(p+1)/2}\right)$$
$$= \frac{p}{1 \times (p-1)} + \frac{p}{2 \times (p-2)} + \cdots + \frac{p}{(p-1)/2 \times (p+1)/2}$$
$$= p\left[\frac{1}{1 \times (p-1)} + \frac{1}{2 \times (p-2)} + \cdots + \frac{1}{(p-1)/2 \times (p+1)/2}\right]$$
$$\text{(mod } p^2\text{)}.$$

Now this, at first, looks like a complication rather than a simplification. But we have gained a very important factor of p on the right-hand side. Now, instead of having to prove that

$$(\text{expression}) = 0 \ (\text{mod } p^2)$$

we now have to prove something like

$$(p \times \text{expression}) = 0 \ (\text{mod } p^2)$$

which is equivalent to proving something of the form

$$(\text{expression}) = 0 \ (\text{mod } p).$$

In other words, we are now reduced to a (mod p) question instead of a (mod p^2) question. Now we have achieved objective (b) given above: reduced the question to that of a smaller modulus, which is well worth the slight increase in complexity.

And it is quickly seen that the apparent increase in expression complexity is just illusionary, as the (mod p) can get rid of a lot more terms than (mod p^2) can. Now, we only have to show that

$$\frac{1}{1 \times (p-1)} + \frac{1}{2 \times (p-2)} + \cdots + \frac{1}{(p-1)/2 \times (p+1)/2} = 0 \ (\text{mod } p).$$

But $p-1$ is equivalent to -1 (mod p), $p-2$ is equivalent to -2 (mod p), and so forth, so the equation reduces to

$$\frac{1}{-1^2} + \frac{1}{-2^2} + \cdots + \frac{1}{-((p-1)/2)^2} = 0 \ (\text{mod } p),$$

or equivalently

$$\frac{1}{1^2} + \frac{1}{2^2} + \frac{1}{3^2} + \cdots + \frac{1}{((p-1)/2)^2} = 0 \ (\text{mod } p).$$

This equation is not too bad, except that the series on the left-hand side ends in an obscure spot (at $1/((p-1)/2)^2$, rather than the more natural $1/(p-1)^2$, for example). But we can 'double up', making use of the fact that $(-a)^2 = a^2$ to get

$$\frac{1}{1^2} + \frac{1}{2^2} + \frac{1}{3^2} + \cdots + \frac{1}{((p-1)/2)^2}$$
$$= \frac{1}{2}\left[\frac{1}{1^2} + \frac{1}{2^2} + \frac{1}{3^2} + \cdots + \frac{1}{((p-1)/2)^2}\right.$$
$$\left. + \frac{1}{(-1)^2} + \frac{1}{(-2)^2} + \frac{1}{(-3)^2} + \cdots + \frac{1}{(-(p-1)/2)^2}\right] \pmod{p}$$
$$= \frac{1}{2}\left[\frac{1}{1^2} + \cdots + \frac{1}{(p-1)^2}\right] \pmod{p}.$$

So proving that $(1/1^2) + \cdots + 1/((p-1)/2)^2$ is equal to 0 (mod p) would be equivalent to proving that $(1/1^2) + \cdots + 1/(p-1)^2$ is equal to 0 (mod p). The latter is more desirable because of its more symmetrical format. (Symmetry is nice to keep—until it can be used to its full effect—while anti-symmetry, is nice to cancel.)

So now we only have to prove

$$\frac{1}{1^2} + \frac{1}{2^2} + \cdots + \frac{1}{(p-1)^2} = 0 \pmod{p}. \tag{8}$$

to prove the whole question. This is tactically a much better formulation than the original one involving numerators and p^2 divisibility, which is a lot stronger (hence harder to prove) than mere p-divisiblity.

So now we have achieved all our tactical goals, and reduced the question down to decent proportions. But where do we go from here? Well, the question seems very closely related to the other problem (7) that we were considering. But we are not going around in circles. Our current goal (8) will imply the original question, whereas (7) was just a side-problem, a simpler version of the question. Rather than going around in circles, we are going around in spirals, heading towards a solution. We have already proved (7): can we prove (8) by the same methods?

Well, we are in luck, because there were two methods we used to solve (7): one was the rearrangement of reciprocals, and the other was cancellation of pairs. Cancellation of pairs unfortunately does not work as well with (8) as it did with (7), mainly because of the squares in the denominators, which produce symmetry rather than anti-symmetry. But the rearrangement method is promising. Take, yet again, the example of $p = 5$ (so we can reuse some previous work):

$$\frac{1}{1^2} + \frac{1}{2^2} + \frac{1}{3^2} + \frac{1}{4^2} = 1^2 + 3^2 + 2^2 + 4^2 \pmod{5}$$
$$= 1^2 + 2^2 + 3^2 + 4^2 \pmod{5}$$
$$= 0.$$

The way it works when $p = 5$ shows the way for the general case. Based on the above examples it looks like the residue classes $1/1, 1/2, 1/3, \ldots, 1/(p-1)$ (mod p) are just a rearrangement of the numbers $1, 2, 3, \ldots, (p-1)$ (mod p); a proof of this fact will be given at the end of this discussion. Thus, we can say that the numbers $1/1^2, 1/2^2, \ldots, 1/(p-1)^2$ are just rearrangements of the numbers $1^2, 2^2, 3^2, \ldots, (p-1)^2$. In other words,

$$\frac{1}{1^2} + \frac{1}{2^2} + \frac{1}{3^2} + \cdots + \frac{1}{(p-1)^2} = 1^2 + 2^2 + 3^2 + \cdots + (p-1)^2 \pmod{p}.$$

This is an easier expression to deal with, because we have removed the reciprocals, which are a nuisance when trying to sum things. In fact, we can now get rid of the sum altogether, using the standard formula

$$1^2 + 2^2 + \cdots + n^2 = \frac{n(n+1)(2n+1)}{6}$$

(which is easily proven by induction), so we have reduced (8) to just proving that

$$\frac{(p-1)p(2p-1)}{6} \equiv 0 \pmod{p}.$$

And one can easily show that this is true when p is a prime greater than 3 (because $(p-1)(2p-1)/6$ is an integer in this case).

So that is it. We keep reducing the equation to simpler and simpler formulations, until it just collapses into nothing. A bit of a long haul, but sometimes it is the only way to resolve these very complicated questions: step-by-step reduction.

Now for the proof that the reciprocals $1/1, 1/2, \ldots, 1/(p-1)$ (mod p) are a permutation of the numbers $1, 2, \ldots, (p-1)$ (mod p): This is equivalent to saying that each non-zero residue (mod p) is the reciprocal of one and only one non-zero residue (mod p), which is obvious.

EXERCISE 2.5. Let $n \geq 2$ be an integer. Show that $1/1 + 1/2 + \cdots + 1/n$ is *not* an integer. (You will need *Bertrand's postulate* (actually a theorem), which shows that given any positive integer n there is at least one prime between n and $2n$.)

EXERCISE 2.6 (*). Let p be a prime and k be a positive integer, not divisible by $p-1$. Show that $1^k + 2^k + 3^k + \cdots + (p-1)^k$ is divisible by p. (Hint: since k could be even, we cannot always use the cancelling trick. However, the rearranging trick will be effective. Let a be a generator of $\mathbb{Z}/p\mathbb{Z}$, so that $a^k \neq 1 \pmod{p}$ when k is not a multiple of $p-1$. Now compute the expression $a^k + (2a)^k + \cdots + ((p-1)a)^k \pmod{p}$ in two different ways.)

3 Examples in algebra and analysis

> One cannot escape the feeling... that these mathematical formulae have an independent existence and an intelligence of their own ... that they are wiser than we are, wiser even than their discoverers ... that we get more out of them that was originally put into them.
> Heinrich Hertz, quoted by F.J. Dyson

Algebra is what most people associate with mathematics. In a sense, this is justified. Mathematics is the study of abstract objects, numerical, logical, or geometrical, that follow a set of several carefully chosen axioms. And basic algebra is about the simplest meaningful thing that can satisfy the above definition of mathematics. There are only a dozen or so postulates, but that is enough to make the system beautifully symmetric. My favourite algebraic identity, to give an example, is

$$1^3 + 2^3 + 3^3 + \cdots + n^3 = (1 + 2 + 3 + \cdots + n)^2.$$

This means, in part, that the sum of the first few cubes will always be a square; for example, $1 + 8 + 27 + 64 + 125 = 225 = 15^2$.

There is more than one algebra, though. Algebra is the study of numbers with the operations of addition, subtraction, multiplication, and division. Matrix algebra, for example, does much the same but with groups of numbers instead of using just one. Other algebras use all kinds of operations and all kinds of 'numbers' but they, sometimes surprisingly, tend to have much of the same properties as normal algebra. For example, a square matrix A can, under special conditions, satisfy the algebraic equation

$$(I - A)^{-1} = I + A + A^2 + A^3 + \cdots.$$

Algebra is the basic foundation of a large part of applied mathematics. Problems of mechanics, economics, chemistry, electronics, optimization, and so on are answered by algebra and differential calculus, which is an advanced form of algebra. In fact, algebra is so important that most of its secrets have been discovered—so it can be safely put into a high-school curriculum. However, a few gems can still be found here and there.

3.1 Analysis of functions

Analysis is also a heavily explored subject, and it is just as general as algebra: essentially, analysis is the study of functions and their properties. The more complicated the properties, the 'higher' the analysis. The lowest form of analysis is studying functions satisfying simple algebraic properties, for instance one can consider a function $f(x)$ such that

$$f \text{ is continuous}, f(0) = 1, \text{ and } f(m+n+1) = f(m) + f(n)$$
$$\text{for all real } m, n \qquad (9)$$

and then deducing properties of the function. For example in this case, there is exactly one function f that obeys the above properties, namely $f(x) = 1 + x$; we would leave this as an exercise. These problems are a good way to learn how to think mathematically, because there is only one or two pieces of data that can be used, so there should be a clear direction in which to go. It is sort of a 'pocket mathematics', where instead of the three dozen axioms and countless thousands of theorems, one only has a handful of 'axioms' (i.e. data) to use. And yet, it still has its surprises.

EXERCISE 3.1. Let f be a function from the reals to the reals obeying (9). Show that $f(x) = 1 + x$ for all real numbers x. (Hint: first prove this for integer x, then for rational x, then finally for real x.)

PROBLEM 3.1 (Greitzer 1978, p. 19).   (*) Suppose f is a function mapping the positive integers to the positive integers, such that f satisfies $f(n+1) > f(f(n))$ for all positive integers n. Show that $f(n) = n$ for all positive integers n.

This equation looks insufficient to prove what we want. After all, how can an inequality prove an equality? Other problems of this type (such as Exercise 3.1) involve functional *equations*, and are easier to handle because one can apply various substitutions and the like and gradually manipulate our original data into a manageable form. This question seems entirely different.

However, if the question is read carefully, we see the function takes integer values, unlike most questions involving functional equations, which usually map onto the real numbers. One immediate way to capitalize on this is to make the inequality 'stronger':

$$f(n+1) \geq f(f(n)) + 1. \qquad (10)$$

Now let us see what we can deduce. The standard method of dealing with these equations is by substituting pertinent values into the variables, so let us start with $n = 1$:

$$f(2) \geq f(f(1)) + 1.$$

This does not tell us much about $f(2)$ or $f(1)$ at first glance, but the $+1$ on the right-hand side hints that the $f(2)$ cannot be too small. In fact as f maps onto the positive integers, $f(f(1))$ must be at least 1, so $f(2)$ is at least 2. Now, we have to show that $f(2)$ is actually 2, so we may be on the right track. (Always try to use tactics that get you closer to the objective, unless all available direct approaches have been exhausted. Only then you should think about going sideways, or—occasionally—backwards.)

So, can we show that $f(3)$ is at least 3? Well, we can try (10) again to get $f(3) \geq f(f(2)) + 1$. By using the same argument as above, we can say that $f(3)$ is at least 2. But can we say something stronger? Earlier we said $f(f(1))$ was at least 1. Perhaps $f(f(2))$ is at least 2. (Indeed, since we 'secretly' know that $f(n)$ should eventually equal n, we know that $f(f(2))$ is 2—but we cannot use that fact yet, since we cannot actually use what we are trying to prove.) With this line of thought one can apply (10) yet again:

$$f(3) \geq f(f(2)) + 1 \geq f(f(2) - 1) + 1 + 1 \geq 3.$$

Here we plugged $f(2) - 1$ into the 'n' of our formula. This works because we already know that $f(2) - 1$ is at least 1.

So it seems we can deduce that $f(n) \geq n$. Because we used the fact that $f(2)$ was at least 2 to prove that $f(3)$ was at least 3, the general proof reeks of induction.

The induction is just a little tricky though. Consider the next case, showing that $f(4) \geq 4$. From (10) we know that $f(4) \geq f(f(3)) + 1$. We already know that $f(3) \geq 3$, so we would like to deduce that $f(f(3)) \geq 3$, in order that we can conclude $f(f(3)) + 1 \geq 4$. To do that, we would like to have in hand a fact of the form 'if $n \geq 3$, then $f(n) \geq 3$'. The easiest way to do that is to put that kind of fact into the induction we are trying to prove. More precisely, we will show:

LEMMA 3.1. $f(m) \geq n$ for all $m \geq n$.

PROOF. We induct on n.

- *Base case* $n = 1$: This is obvious: we are given that $f(m)$ is a positive integer, hence $f(m)$ is at least 1.
- *Induction case*: Assume that the lemma works for n, and we will try to prove $f(m) \geq n + 1$ for all $m \geq n + 1$. Well, for any $m \geq n + 1$,

we can use (10) to obtain $f(m) \geq f(f(m-1)) + 1$. Now $(m-1) \geq n$, hence $f(m-1) \geq n$ (by induction hypothesis). We can go further: since $f(m-1) \geq n$, then by the induction hypothesis again $f(f(m-1)) \geq n$. Therefore, $f(m) \geq f(f(m-1)) + 1 \geq n+1$, and the induction hypothesis is proved. □

If we specialize Lemma 3.1 to the case $m = n$, we obtain our subgoal:

$$f(n) \geq n \quad \text{for all positive integers } n. \tag{11}$$

Now what? Well, as with all functional equation questions, once we have a new result, we should just play around with it and try to recombine it with previous results. Our only previous result is (10), so we can put our new equation into (10). The only useful result we get is

$$f(n+1) \geq f(f(n)) + 1 \geq f(n) + 1$$

which follows once we replace n by $f(n)$ in (11). In other words,

$$f(n+1) > f(n).$$

This is a very useful formula: this means that f is an increasing function! (not obvious from (10), is it?) This means that $f(m) > f(n)$ if and only if $m > n$. This means our original equation

$$f(n+1) > f(f(n))$$

could be reformulated as

$$n + 1 > f(n).$$

And this, with (11), proves what we wanted.

PROBLEM 3.2 (Australian Mathematics Competition 1984, p. 7).
Suppose f is a function on the positive integers which takes integer values with the following properties:

(a) $f(2) = 2$
(b) $f(mn) = f(m)f(n)$ for all positive integers m and n
(c) $f(m) > f(n)$ if $m > n$.

Find $f(1983)$ (with reasons, of course).

Now we have to find out a particular value of f. The best way is to try to evaluate all of f, not just $f(1983)$. (1983 is just the year of the question anyway.) This is, of course, assuming there is only one solution of f. But implicit in the question is the fact that there is only one possible value of $f(1983)$ (otherwise there would be more than one answer), and because of the ordinariness of 1983 we might reasonably conjecture that there is only one solution to f.

So, what are the properties of f? We know that $f(2) = 2$. Repeated application of (b) yields $f(4) = f(2)f(2) = 4$, $f(8) = f(4)f(2) = 8$, etc. Indeed, an easy induction shows that $f(2^n) = 2^n$ for all n. So $f(x) = x$ when x is a power of 2. Perhaps $f(x) = x$ for all x. Plugging this back into (a), (b), and (c) shows that this works: $f(x) = x$ is one solution of (a), (b), and (c). So, if we think that there is only one solution of f, then this one has to be it. So we might want to prove the more general, but clearer question:

> The only function from the positive integers to the integers satisfying (a), (b), and (c) is the identity function (i.e. $f(n) = n$ for all n).

So we have to prove that if f satisfies (a), (b), and (c), then $f(1) = 1$, $f(2) = 2$, $f(3) = 3$, and so on. Let us first try to prove $f(1) = 1$ (with functional equations we should try small examples first to get a 'feel' of the question). Well, by (c) we know that $f(1) < f(2)$, and we know $f(2)$ is 2, so $f(1)$ is less than 2. And by (b), we get (with $n = 1$ and $m = 2$)

$$f(2) = f(1)f(2),$$

and thus

$$2 = 2f(1).$$

This means $f(1)$ must equal 1, as desired.

We now have $f(1) = 1$ and $f(2) = 2$. What about $f(3)$? (a) is no help, and (b) only gives $f(3)$ in terms of other numbers like $f(6)$ or $f(9)$, which is also of not much help. (c) yields

$$f(2) < f(3) < f(4)$$

but $f(2)$ is 2 and $f(4)$ is 4, so

$$2 < f(3) < 4.$$

But the only integer between 2 and 4 is 3. Hence $f(3)$ must be 3.

This gives us a clue: $f(3)$ was 3 only because it was an integer (see how this is similar to the previous question, $f(n+1) > f(f(n))$?). Without this restriction, $f(3)$ could have been 2.1, or 3.5, or whatever. Let us see if we can use this clue more often.

We already know $f(4) = 4$; let us try to work out $f(5)$. Using (c) in the hope of doing what we did to $f(3)$, we get

$$f(4) < f(5) < f(6).$$

Now $f(4)$ is 4. But what about $f(6)$? Never fear: 6 is 2 times 3, so $f(6) = f(2)f(3) = 2 \times 3 = 6$. Therefore, $f(5)$ is between 4 and 6, and must be 5. This seems to be going well; we have now worked out all the values of $f(n)$ up to $n = 6$.

Because we seem to be relying on past results to attain the new ones, the general proof smells heavily on induction. And because we are not just using one previous result, but several previous results, we probably need *strong* induction.

LEMMA 3.2. $f(n) = n$ for all n.

PROOF. We use strong induction. First we check the base case: does $f(1) = 1$? Yes, we have already shown this. Now assume that $m \geq 2$, and that $f(n) = n$ for all n smaller than m. We want to show $f(m) = m$. Looking at a few examples we will soon see that we have to divide into cases: m even and m odd.

Case 1: m is even. In this case we can write $m = 2n$ for some integer n. n is less than m, so by strong induction hypothesis $f(n) = n$. Hence $f(m) = f(2n) = f(2)f(n) = 2n = m$, as desired.

Case 2: m is odd. Here we write $m = 2n + 1$. By (c), $f(2n) < f(m) < f(2n+2)$. By strong induction $f(2n) = 2n$ and $f(n+1) = n+1$ because both $n+1$ and $2n$ are smaller than m. Now by (b) $f(2n+2) = f(2)f(n+1) = 2(n+1) = 2n+2$, so our inequality becomes

$$2n < f(m) < 2n+2,$$

and so $f(m) = 2n + 1 = m$, as desired. So either way, the induction hypothesis holds. □

And so by strong induction $f(n)$ is forced to be n. So to answer our question $f(1983)$ must be 1983, and that's that.

EXERCISE 3.2. Show that Problem 3.2 can still be solved if we replace (a) with the weaker condition

(a') $f(n) = n$ for at least one integer $n \geq 2$.

EXERCISE 3.3. (*) Show that Problem 3.2 can still be solved if we allow $f(n)$ to be a real number, rather than just an integer. (Hint: first try to prove that $f(3) = 3$, by comparing $f(2^n)$ with $f(3^m)$ for various integers n, m.) For an additional challenge, solve Problem 3.2 with this assumption *and* with (a) replaced by (a').

EXERCISE 3.4 (1986 International Mathematical Olympiad, Q5.). (**) Find all (if any) functions f taking the non-negative reals onto the non-negative reals, such that

(a) $f(xf(y))f(y) = f(x+y)$ for all non-negative x, y;
(b) $f(2) = 0$;
(c) $f(x) \neq 0$ for every $0 \leq x < 2$.

(Hint: The first condition involves products of values of the function, and the other two conditions involve a function having a value of zero (or non-zero). Now, what can one say when a product equals 0?)

3.2 Polynomials

Many algebra questions concern polynomials of one or more variables, so let us pause for a bit to recall some definitions and results concerning these polynomials.

A *polynomial of one variable* is a function, say $f(x)$, of the form

$$f(x) = a_n x^n + a_{n-1} x^{n-1} + a_{n-2} x^{n-2} + \cdots + a_1 x + a_0$$

or, to be more formal,

$$f(x) = \sum_{i=0}^{n} a_i x^i.$$

The a_is are constants (in this book they will always be real numbers), and it is assumed that a_n is not zero. We call n the *degree* of f.

Polynomials in many variables, say three variables for the sake of example, do not have as nice a form as the one-dimensional polynomials, but are quite useful nevertheless. Anyway, $f(x,y,z)$ is a *polynomial in three variables* if it takes the form

$$f(x,y,z) = \sum_{k,l,m} a_{k,l,m} x^k y^l z^m,$$

where the $a_{k,l,m}$ are (real) constants, and the summation runs over all non-negative k, l, and m such that $k+l+m \leq n$, and it is assumed that at least one of the non-zero $a_{k,l,m}$ satisfy $k+l+m = n$. We again call n the *degree* of f; polynomials of degree 2 are *quadratic*, degree 3 are *cubic*, and so forth. If the degree is 0, then the polynomial is said to be *trivial* or *constant*. If all non-zero $a_{k,l,m}$ satisfy $n = k+l+m$, then f is said to be *homogeneous*. Homogeneous polynomials have the property that

$$f(tx_1, tx_2, \ldots, tx_m) = t^m f(x_1, x_2, \ldots, x_m)$$

for all $x_1, \ldots, x_m, t$. For example, $x^2 y + z^3 + xz$ is a polynomial of three variables $(x, y, \text{and } z)$ and has degree 3. It is not homogeneous because the xz term has a degree of 2.

A polynomial f of m variables is said to be *factored* into two polynomials p and q if $f(x_1, \ldots, x_m) = p(x_1, \ldots, x_m) q(x_1, \ldots, x_m)$ for all $x_1, \ldots, x_m$; p and q are then said to be *factors* of f. It is easily proven that the degree of a polynomial is equal to the sum of the degrees of the factors. A polynomial is *irreducible* if it cannot be factored into non-trivial factors.

The *roots* of a polynomial $f(x_1, \ldots, x_m)$ are the values of $(x_1, \ldots, x_m)$ which return a zero value, so that $f(x_1, \ldots, x_m) = 0$. Polynomials of one variable can have as many roots as their degree; in fact, if multiplicities and complex roots are counted, polynomials of one variable always have exactly as many roots as their degree. For instance, the roots of a quadratic polynomial $f(x) = ax^2 + bx + c$ is given by the well-known quadratic formula

$$x = \frac{-b \pm \sqrt{b^2 - 4ac}}{2a}.$$

Cubic and quartic polynomials also have formulae for their roots, but they are much messier and not very useful in practice. Once one gets to quintic and higher polynomials, there is no elementary formula at all! And polynomials of two or more variables are even worse; typically one has an infinite number of roots.

The roots of a factor are a subset of the roots of the original polynomial; this can be a useful piece of information in deciding whether one polynomial divides another. In particular, $x - a$ divides $f(x)$ if and only if $f(a) = 0$, because a is a root of $x - a$. In particular, for any polynomial $f(x)$ of one variable and any real number t, $x - t$ divides $f(x) - f(t)$.

Now let us tackle some questions involving polynomials.

PROBLEM 3.3 (Australian Mathematics Competition 1987, p. 13). Let a, b, c be real numbers such that

$$\frac{1}{a} + \frac{1}{b} + \frac{1}{c} = \frac{1}{a+b+c} \tag{12}$$

with all denominators non-zero. Prove that

$$\frac{1}{a^5} + \frac{1}{b^5} + \frac{1}{c^5} = \frac{1}{(a+b+c)^5}. \tag{13}$$

At first this question looks simple. There is really only one piece of information given, so there should be a straightforward sequence of logical steps leading to the result we want. Well, an initial attempt to deduce the second equation from the first may be to raise both sides of (12) to the fifth power, which gets something similar to the desired result, but with a whole lot of messy terms on the left-hand side. There seems to be no other obvious manipulation. So much for the direct approach.

On second glance, the first equation looks suspect, like one of the equations that high-school students are warned not to use because they are usually fallacious. This gives us our first real clue: the first equation should restrict a, b, and c quite a bit. It may be worth reinterpreting the equation (12).

A common denominator seems like a good start. Combining the three reciprocals on the left-hand side we get

$$\frac{ab + bc + ca}{abc} = \frac{1}{a+b+c}$$

and cross-multiplying we get

$$ab^2 + a^2b + a^2c + ac^2 + b^2c + bc^2 + 3abc = abc. \tag{14}$$

At this point one may think of the various inequalities one could use here: Cauchy-Schwarz, arithmetic mean-geometric mean, etc. (Hardy 1975, pp. 33–34). That would not be so bad if a, b, and c were constrained to be positive, but there is no such constraint: in fact the condition cannot hold if a, b, and c are positive as $1/(a+b+c)$ would then be smaller than all three reciprocals on the left-hand side of (12).

Since (14) is equivalent to (12), and is algebraically simpler ((14) contains no reciprocals), we could try to deduce (13) from (14). Again, the direct approach is not feasible. Usually the only other way to deduce an equation from some others are by proving an intermediate result, or by doing some useful substitution. (There are more exotic alternatives, like considering (12) as a contour of the function $(1/a) + (1/b) + (1/c) - (1/a+b+c)$ and then using calculus to find the shape and properties of that contour, but it is best to stick with the easy options first.)

Substitutions do not seem appropriate: the equations (12) or (14) are simple enough as they are, and substitutions would not make them much simpler. So we will try to guess and prove an intermediate result. The best kind of intermediate result is a parameterization, as this can be substituted directly into the desired equation. One way of parameterizing is to solve for one of the variables, say a. The equation (14) will not solve for a easily (unless you are willing to use the quadratic formula). The equation (12) *does* solve for a, and you can prove our question by solving for a, b, and c in turn and deducing an intermediate result (which incidentally is equivalent to the result I will give below. But it would have to be, would not it?). But I will try something else.

Failing a parameterization, one could simply recast (14) into a better form. Solutions of (14) are essentially the roots of the polynomial $a^2b+b^2a+b^2c+c^2b+c^2a+a^2c+2abc$. The best way to deal with roots of polynomials is to factorize the polynomial (and vice versa). What are the factors? Because we know that (14) must somehow imply (13), we should be pretty confident that there is some workable form of (14) that will lead to $5ab$, and the only workable form of a polynomial is a breakup into factors. But to find out what they are, we have to experiment. The polynomial is homogeneous, so its factors should be too. The polynomial is symmetric, so the factors should be symmetries of each other. The polynomial is cubic, so there should be a linear factor. We should now try factors of the form $a+b$, $a-b$, a, $a+b+c$, $a+b-c$, and so on. (Things like $a+2b$ might also work, but are not as 'nice' and in any case can be tried afterwards.) It is soon apparent (from the factor theorem) that $a+b$, and similarly $b+c$ and $c+a$ are roots of the cubic. From there it is easy to verify that (14) is factorizable into $(a+b)(b+c)(c+a)$. This means that (12) is true if and only if either $a+b=0$, $b+c=0$, or $c+a=0$. Substituting each of these possibilities into $5ab$ does the trick.

EXERCISE 3.5. Factorize $a^3 + b^3 + c^3 - 3abc$.

EXERCISE 3.6. Find all integers a, b, c, d such that $a+b+c+d = 0$ and $a^3 + b^3 + c^3 + d^3 = 24$. (Hint: it is not hard to guess *some* solutions to these equations, but to show that you have *all* of them, substitute the first equation into the second and factorize.)

The factorization of polynomials, or impossibility thereof, is a fascinating piece of mathematics. The following question is instructive because it uses just about every trick in the book to find a solution.

PROBLEM 3.4. (**) Prove that any polynomial of the form $f(x) = (x - a_0)^2(x - a_1)^2, \ldots, (x - a_n)^2 + 1$ where $a_0, a_1, \ldots, a_n$ are all integers, cannot be factorized into two non-trivial polynomials, each with integer coefficients.

This is a rather general statement: it says for example, that the polynomial

$$(x - 1)^2(x + 2)^2 + 1 = x^4 + 2x^3 - 3x^2 - 4x + 5$$

cannot be factorized into other integer polynomials. How can we prove that?

Well, suppose that $f(x)$ is factorizable into two non-trivial integer polynomials, $p(x)$ and $q(x)$. Then $f(x) = p(x)q(x)$ for all x. Big deal. But remember that f has this special property: it is some sort of square plus one. How can we use this? Well, we could say that $f(x)$ is always positive (or even that $f(x) \geq 1$), but that does not say much about $p(x)$ and $q(x)$ except that they are of the same sign. However, we have another piece of data; f is not just any old square plus one; the square is a square of a combination of linear factors. Can we use these $(x - a_i)$'s to our advantage?

Well the nicest factor one can have is 0, because that makes the entire expression 0. (Actually, there are also occasions where having a 0 factor is the last thing one wants to have, because one may wish to cancel this factor.) $(x - a_i)$ is 0 when, well, x is a_i, so here we have an idea: plug in a_i for x. We get

$$f(a_i) = \ldots (a_i - a_i)^2 \cdots + 1 = 1.$$

Getting back to $p(x)$ and $q(x)$, this result means that

$$p(a_i)q(a_i) = 1.$$

What does this mean? Very little, unless one remembers that p and q have integer coefficients, and that the a_i are integers too. The upshot of this is that $p(a_i)$ and $q(a_i)$ are both integers. So we have two integers multiplying to 1. This can only occur when the integers are either both 1, or both -1. In shorthand,

$$p(a_i) = q(a_i) = \pm 1 \quad \text{for all } i = 0, 1, \ldots, n.$$

One should be a little careful with the $\pm$ notation here; we know that $p(a_1)$ and $q(a_1)$, for instance, are equal to each other, but $p(a_1)$ and $p(a_2)$ could have the same sign or the opposite sign, for all we know right now.

We have found, more or less, the value of $p(a_0), \ldots, p(a_n)$ and $q(a_0), \ldots, q(a_n)$, so each polynomial is 'pegged' by n points. But polynomials only have as many degrees of freedom as their degree. Now $pq = f$, so the degree of p plus the degree of q equals the degree of f, which is $2n$. This means that one of the polynomials, say p, has a degree of at most n. In summary, we have a polynomial with degree at most n but restricted to lie on n given points. Hopefully this can be exploited to a contradiction, which is what we are searching for.

What do we know about a polynomial that has degree at most n? Well, it has at most n roots. Do we know anything about the roots of p? Well, p is a factor of f, so the roots of p are also roots of f. What are the roots of f? There are none! (Well, none on the real line, at least.) f is always positive (in fact, it is always at least 1), and hence can have no roots. This means in turn that p can have no roots. What does it mean when a polynomial has no roots? It means that it never crosses 0, that is, it never changes sign. In other words, p is either always positive or always negative. This gives us two cases, but we can save a little bit of work by observing that one case implies the other. Indeed, if we have one factorization $f(x) = p(x)q(x)$, we automatically have another factorization $f(x) = (-p(x))(-q(x))$. So if p is always negative, we can always flip the factorization and end up with a new factorization where p is always positive.

So without loss of generality we will take p to be always positive. We already know that $p(a_i)$ is either $+1$ or -1, and now we know it is positive as well, so $p(a_i)$ has to be $+1$ for all i. And $q(a_i)$ is forced to be equal to $p(a_i)$, so $q(a_i)$ is also $+1$ for all i. Now what?

Well, $p(x)$ and $q(x)$ are forced to take on the value of $+1$ at least n times. This can be rephrased in terms of roots, as follows: $p(x) - 1$ and $q(x) - 1$ have at least n roots. But $p(x) - 1$ has a degree of at most n, because $p(x)$ itself has a degree of at most n. This means that the only way $p(x) - 1$ can have n roots is if $p(x) - 1$ has degree exactly n. This in turn means that $p(x)$ is of degree n, and hence $q(x)$ is also of degree n.

To sum up what we know so far: We have assumed that $f(x) = p(x)q(x)$. p and q are both positive integer polynomials of degree n, and $p(a_i) = q(a_i) = 1$, or alternatively $p(a_i) - 1 = q(a_i) - 1 = 0$, for all i. Now we know the roots of $p(x) - 1$: they are the a_is. They are the only roots of $p(x) - 1$, because $p(x) - 1$ can only have n roots at most. This means that

$p(x) - 1$ is of the form

$$p(x) - 1 = r(x - a_1)(x - a_2)\ldots(x - a_n)$$

and similarly $q(x) - 1$ is of the form

$$q(x) - 1 = s(x - a_1)(x - a_2)\ldots(x - a_n),$$

where r and s are some constants. To find out more about r and s, remember that p and q are integer polynomials. The leading coefficient of $p(x) - 1$ is r, and the leading coefficient of $q(x) - 1$ is s. This means that r and s have to be integers.

Now we apply these formulas for $p(x)$ and $q(x)$ into our original formula $f(x) = p(x)q(x)$ and get

$$(x - a_1)^2(x - a_2)^2 \ldots (x - a_n)^2 + 1 = (r(x - a_1)(x - a_2)\ldots(x - a_n) + 1)$$
$$\times (s(x - a_1)(x - a_2)\ldots(x - a_n) + 1).$$

This equation compares two explicitly defined polynomials. The best thing to do now is to compare coefficients.

Comparing the x^n coefficients we get $1 = rs$ and because r and s are integers, this means that either $r = s = +1$ or $r = s = -1$. Let us first suppose that $r = s = 1$. Our polynomial equation is now

$$(x - a_1)^2(x - a_2)^2 \ldots (x - a_n)^2 + 1 = ((x - a_1)(x - a_2)\ldots(x - a_n) + 1)$$
$$((x - a_1)(x - a_2)\ldots(x - a_n) + 1).$$

Upon expanding and cancelling, this becomes

$$2(x - a_1)(x - a_2)\ldots(x - a_n) = 0$$

which is ridiculous (it must hold for all x). The case $r = s = -1$ is similar, and we are done.

EXERCISE 3.7. Prove that the polynomial $f(x) = (x - a_1)(x - a_2)\ldots(x - a_n) + 1$ cannot be factorized into two smaller integer polynomials, where the a_is are integers. (Hint: if $f(x)$ factors into two polynomials $p(x)$ and $q(x)$, look at $p(x) - q(x)$. Note that this particular strategy could also be applied to Problem 3.4, but turns out to be somewhat ineffective in that case.)

EXERCISE 3.8. Let $f(x)$ be a polynomial with integer coefficients, and let a, b be integers. Show that $f(a) - f(b)$ can only equal 1 when a, b are consecutive. (Hint: factorize $f(a) - f(b)$.)

4 Euclidean geometry

> Archimedes will be remembered when Aeschylus is forgotten, because languages die and mathematical ideas do not.
> G.H. Hardy, 'A Mathematicians Apology'

Euclidean geometry was the first branch of mathematics to be treated in anything like the modern fashion (with postulates, definitions, theorems, and so forth); and even now geometry is conducted in a very logical, tightly knit fashion. There are several basic results which can be used to systematically attack and resolve questions about geometrical objects and ideas. This idea can be taken to extremes with coordinate geometry, which transforms points, lines, triangles, and circles into a quadratic mess of coordinates, crudely converting geometry into algebra. But the true beauty of geometry is in how a non-obvious looking fact can be shown to be undeniably true by the repeated application of obvious facts. Take, for example, Thales' theorem (Euclid III, 31):

THEOREM 4.1 (Thales' theorem). The angle subtended by a diameter is a right angle. In other words, in the diagram below, we have $\angle APB = 90°$.

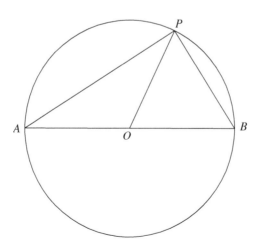

PROOF. If we draw the line segment OP, we have divided the triangle into two isosceles triangles (because $|OP| = |OA|$ and $|OP| = |OB|$; here we use $|AB|$ to denote the length of the line segment AB). Now using the facts that isosceles triangles have equal base angles, and that the sum of the angles in a triangle is $180°$, we have

$$\angle APB = \angle APO + \angle OPB = \angle PAO + \angle PBO = \angle PAB + \angle PBA$$
$$= 180° - \angle APB,$$

and hence APB must be a right angle. □

Geometry is full of things like this: results you can check by drawing a picture and measuring angles and lengths, but are not immediately obvious, like the theorem that the midpoints of the four sides of a quadrilateral always make up a parallelogram. These facts—they have a certain something about them.

> PROBLEM 4.1 (Australian Mathematics Competition 1987, p. 12).
> ABC is a triangle that is inscribed in a circle. The angle bisectors of A, B, C meet the circle at D, E, F, respectively. Show that AD is perpendicular to EF.

The first step, of course, is to draw a picture and label what we can:

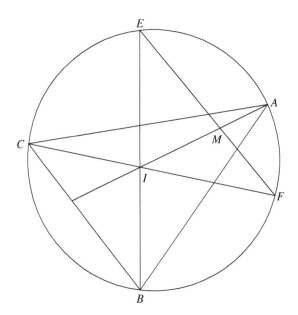

I have taken the liberty of labelling the incentre I (the intersection of all the bisectors, and likely to be important), as well as the intersection M of AD and EF (which is where we want to establish our right-angle).

Thus, we can write our objective now as an equation: we want to show that $\angle AMF = 90°$.

This is a feasible-looking problem: the diagram is easy to draw, the conclusion is quite evident from the figure. With such a problem, a direct approach would probably work quite nicely.

We need to compute an angle at M. At first sight, the point M is rather unremarkable. But after filling in some data we see that we already have a wealth of other angles, mainly due to all the angle bisectors, and triangles and circles around the place. Perhaps just by finding enough angles we can determine $\angle AMF$. After all, there are heaps of theorems just waiting to be used: the sum of angles in a triangle add up to $180°$; the angle subtended by a chord on an arc is always the same; the angle bisectors are concurrent.

We need some angles to start with. With the 'main' triangle being ABC, and with all the angle bisectors and circles and stuff revolving about this triangle, it might be best to start with the angles $\alpha = \angle BAC$, $\beta = \angle ABC$, $\gamma = \angle BCA$ (it is traditional to use Greek letters to denote angles). Of course, we have $\alpha + \beta + \gamma = 180°$. We can fill in lots of other angles, too: for example, $\angle CAD = \alpha/2$. (It is best if you draw a sketch of the figure and fill in the angles yourself.) Then, we can make use of the fact that the angles of the triangle add up to $180°$, and work out some of the inner angles. For example, if I is the incentre of ABC (the intersection of AD, BE, and CF), then we can easily say that $\angle AIC = 180° - \alpha/2 - \gamma/2$, by considering the triangle AIC. In fact we can get virtually all relevant angles—except those at M, which are the ones we really want. So, we must somehow represent our angle at M in terms of angles that are not related to M. Well, this is easily done; we could say, for example, that $\angle IMF$, which we want to be $90°$, can be written as

$$\angle IMF = 180° - \angle MIF - \angle IFM = 180° - \angle AIF - \angle CFE.$$

This is progress because the two angles $\angle AIF$ and $\angle CFE$ are much more easily worked out. Indeed we have

$$\angle AIF = 180° - \angle AIC = \alpha/2 + \gamma/2,$$

and because equal chords subtend equal angles we have

$$\angle CFE = \angle CBE = \beta/2.$$

Hence

$$\angle IMF = 180° - \alpha/2 - \beta/2 - \gamma/2 = 180° - 180°/2 = 90°,$$

as desired.

This is a lovely way to solve some geometrical questions: by simply working out angles. They are usually easier to work out than sides (which have all kinds of nasty sine and cosine rules to plough through), and the rules are easier to remember. They are best for questions which have no reference to side lengths, and have lots of triangles and circles to play with, and maybe even an isosceles triangle to work with. But to get some of the more obscure angles, you usually have to work out a heck of a lot of other angles first.

> PROBLEM 4.2 (Taylor 1989, p. 8, Q1). In triangle BAC the bisector of the angle at B meets AC at D; the angle bisector of C meets AB at E. These bisectors meet at O. Suppose that $|OD| = |OE|$. Prove that either $\angle BAC = 60°$ or that BAC is isosceles (or both).

We should first draw a picture. It is a little tricky because one has to somehow make OD and OE the same length, but we can cheat a bit by making ABC isosceles or $\angle BAC = 60°$ (since we know that is what is supposed to happen anyway). This gives two possible configurations:

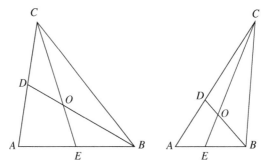

We have only one piece of data: $|OD| = |OE|$, and we wish to prove a strange looking result: a choice of two properties about our triangle. But both properties are angle-related (isosceles triangles have equal base angles, and angle bisectors are clearly angle-related), so one can assume this to be an angle problem (at first, anyway).

Once we have decided to attack the problem with angles, it remains to rephrase the given data $|OD| = |OE|$ in terms of angles. The obvious way is to say that, because ODE is isosceles, $\angle ODE = \angle OED$. That may look promising, but it is quite hard to equate the angles $\angle OED$ and $\angle ODE$ to any other angles. In particular, we would like those angles to be in terms of angles $\alpha = \angle BAC$, $\beta = \angle ABC$, $\gamma = \angle ACB$ because we want to prove that either $\beta = 60°$ or $\alpha = \gamma$. (Also, ABC is the 'main' triangle: all other points stem from that triangle. It is a logical reference frame; all quantities should be in terms of the main triangle.) But there are other ways to translate sides into angles.

Let us look at OD and OE. We want to relate these lengths to our angles α, β, γ. There are several ways to relate sides and angles: basic trigonometry, similar triangles, isosceles and equilateral triangles, sine and cosine rules, to mention a few. Basic trigonometry requires right angles and circles, and we do not have many of them. We have few similar triangles either, and we have already considered the isosceles triangle approach. The cosine rule usually complicates rather than simplifies, and it just creates more unknown lengths. This only leaves the sine rule as a feasible alternative. After all, it relates sides to angles quite directly.

Well, to use the sine rule, we need a triangle or two, preferably those that contain OD and OE and have a lot of angles that we already know. Looking at the diagram, and measuring out angles, we can guess that the triangles AOD, COD, AOE, BOE could be useful. The triangles AOE and AOD have a common side, which should make the problem simpler, so we should try these triangles first (always try to look for connections. Knowing that two quantities are equal would not help unless you connect them in some way). Since we are only looking at four of the six points (A,D,E,O) we could draw a reduced diagram to deal with them (after all, why should one have to deal with useless clutter?).

We know that $\angle EAO = \angle DAO = \alpha/2$, and by the fact that the sum of the angles in a triangle is $180°$ we can work out $\angle AEO = 180° - \alpha - \gamma/2 = \beta + \gamma/2$, and similarly we have $\angle ADO = 180° - \alpha - \beta/2 = \gamma + \beta/2$. One can also fill in a couple more angles connecting A, D, E, O, and eventually our reduced diagram looks something like this (rotated and blown up for clarity):

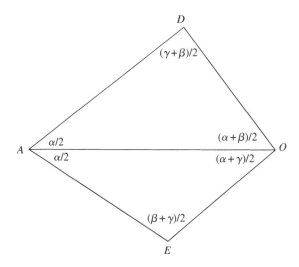

Now we can use the sine rule. To get a workable expression for $|OD|$ and $|OE|$ (which was why we wanted the sine rule in the first place), we can say

$$\frac{|OD|}{\sin(\alpha/2)} = \frac{|OA|}{\sin(\gamma + (\beta/2))} = \frac{|DA|}{\sin(\alpha/2 + \beta/2)}$$

and

$$\frac{|OE|}{\sin(\alpha/2)} = \frac{|OA|}{\sin(\beta + (\gamma/2))} = \frac{|EA|}{\sin(\alpha/2 + \gamma/2)}.$$

Now that we have an equation, our given data ($|OD| = |OE|$) may resolve into something useful. The length $|OA|$ appears in both of the above sets of equations, so we should perhaps put $|OD|$ and $|OE|$ in terms of $|OA|$. We get

$$|OD| = |OA| \frac{\sin(\alpha/2)}{\sin(\gamma + \beta/2)}$$

$$|OE| = |OA| \frac{\sin(\alpha/2)}{\sin(\beta + \gamma/2)}$$

and so $|OD| = |OE|$ if and only if $\sin(\gamma + \beta/2) = \sin(\beta + \gamma/2)$. (Actually, something silly could happen such as $\sin(\alpha/2) = 0$. It does not take long to see that these situations only exist in extremely degenerate cases, and these freak cases are easily dealt with separately. But always remember to watch out for these things.)

We have now converted an equality about edges into an equality about angles. More importantly, these angles are relevant to our objective (which involves the angles α, β, γ) so we must be heading in the right direction. The question is now almost completely algebraical.

Anyway, two sines are equal. This can mean two things:

$$\gamma + \beta/2 = \beta + \gamma/2$$

or

$$\gamma + \beta/2 = 180° - (\beta + \gamma/2).$$

We are getting closer and closer to our objective; the sines are gone, and we have also for the first time gotten a statement that involves an 'or'. Now it is not hard to see that the first case leads to $\beta = \gamma$, while the second leads to $\beta + \gamma = 120°$, and hence $\alpha = 60°$. And we have managed, strangely enough, to stumble onto our objective.

And there we have it. Sometimes, we can just leap on our given data and hammering it into an equation resembling our objective (in this case, anything involving α, β, and γ) and then applying simple algebra to change it to what we want. This is called the *direct* or *forward* approach, and it works well when the objective is a simple relation involving easily calculable parts of the problem, because then we can have an idea how to approach the problem, by gradually simplifying and transforming our data into things that look more and more like our objective. When the objective is obscure, we may have to transform the objective before we know which directions to try, as the next problem demonstrates.

PROBLEM 4.3 (Australian Mathematics Competition 1987, p. 13). (*) Let $ABFE$ be a rectangle and D be the intersection of the diagonals AF and BE. A straight line through E meets the extended line AB at G and the extended line FB at C so that $|DC| = |DG|$. Show that $|AB|/|FC| = |FC|/|GA| = |GA|/|AE|$.

With geometry problems, one can either work forward (measuring sides and angles systematically), or backward (turning the end result into something equivalent but simpler to work with). Simply drawing a figure and guessing conclusions is helpful sometimes, but a figure is quite hard to draw in this case. How do you force it so that $|DC| = |DG|$? A bit of trial and error (and taking a peek at the conclusion $|AB|/|FC| = |FC|/|GA| = |GA|/|AE|$ eventually allows one to draw a decent picture:

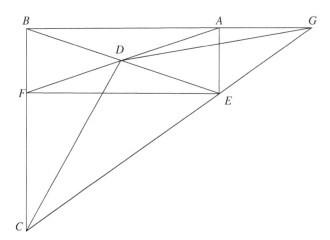

Let us try the forward approach. Hack-and-slash coordinate geometry is one long and boring way that is prone to abysmal complications and huge errors. Let us try that as a last resort (although the right angle at A looks like a tempting location to place the origin and coordinate axes). Vector geometry is also not suited too well with equations like $|DC| = |DG|$ (but the vector versions are still usually neater than the coordinate counterparts).

How about measuring lines and angles? We only know that we have four right angles on the rectangle, and we also know $|DC| = |DG|$. So, maybe DCG is isosceles, but that does not say much. Dropping perpendiculars from D to CG or doing other similar constructions are not much help either. (As we shall see later on, a certain construction does help, but it is definitely not intuitively obvious in the forward approach.)

Backwards we go then. We want to prove that three ratios are equal to each other. This suggests similar triangles. Can we form a triangle out of, say, AB and FC? Not quite: but we can form a triangle out of FE and FC, and FE is equal in length to AB. Once we recognize one triangle, the other two should not be too hard. Just from looking at triangle FCE in our diagram we can see (and easily prove) that it is similar to BCG and AEG, so

$$|EF|/|FC| = |GB|/|BC| = |GA|/|AE|.$$

And, keeping our goal in mind, we convert that to

$$|AB|/|FC| = |GB|/|BC| = |GA|/|AE|. \qquad (15)$$

So we have already proved that two of the three ratios we require, $|AB|/|FC|$ and $|GA|/|AE|$, are equal to each other. The third ratio that we want, $|FC|/|GA|$, cannot be easily converted into a triangle. But look at the middle ratio in (15). The pair of edges look vaguely related to FC and GA. In fact FC is a segment of BC and GA is a segment of BG. This hints that it might be easier to prove

$$|FC|/|GA| = |GB|/|BC|$$

than

$$|AB|/|FC| = |FC|/|GA| \text{ or } |FC|/|GA| = |GA|/|AE|.$$

Besides, this formulation is more symmetric and involves only one equality.

Even with our 'possibly simpler' formulation, there still seems to be no similar triangles to use. At this point we need to manipulate our problem further. One obvious thing to try is to rearrange the ratios, perhaps by multiplying them out to get

$$|FC| \times |BC| = |AG| \times |BG|$$

or by switching the ratios and getting

$$|FC|/|BG| = |GA|/|BC|.$$

This does not seem to be much of an improvement. But the terms $|FC| \times |BC|$ and $|AG| \times |BG|$ might look a little familiar. In fact one might be reminded of the following result (which is usually in high-school textbooks but rarely used there):

THEOREM 4.2. If P is outside a circle with centre O and radius r, and a ray from P cuts the circle at two points Q, R, then

$$|PQ| \times |PR| = |PT|^2 = |PO|^2 - r^2,$$

where T is a point where one of the two tangents from P meets the circle.

PROOF. Observe that PQT is similar to PTR, so that $|PQ|/|PT| = |PT|/|PR|$. Also, from Pythagoras' theorem we have $|PO|^2 = |PT|^2 + r^2$, and the claims follow. □

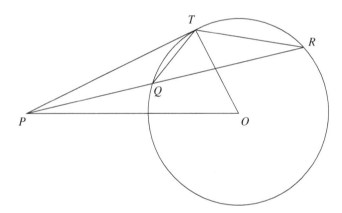

To use Theorem 4.2, we first need to create a circle. We want to evaluate $|FC| \times |BC|$ and $|AG| \times |BG|$. The circle must hence include the points F, B, and A. Now it just so happens that the circle touching the points F, B, and A has D as its centre (Theorem 4.1!). So by Theorem 4.2 we have

$$|FC| \times |BC| = |DC|^2 - r^2$$

and

$$|AG| \times |BG| = |DG|^2 - r^2,$$

where r is the radius of the circle. And as we are conveniently given the fact that $|DC| = |DG|$, our result is proved.

This is the way pure geometry questions work: with seemingly barely enough data to go on, and something quite obscure to prove, one usually needs a special way to do it. A construction or something might make it clear. Look for things that vaguely trigger a memory. For example, if in some geometry problem you had to prove $\angle ABC = \angle ADC$, you could prove that $ABDC$ is cyclic instead, which is equivalent (if B, D are on the same side of AB). Or if you had to prove $|AB| > |AC|$, you could equivalently prove that $\angle ACB > \angle ABC$ (provided A, B, C are not collinear). Or if a question about areas of various triangles crops up, use facts like triangles with equal base and equal height have equal area, or perhaps if the base of a triangle was halved, then the area would also be halved. This does not mean that you should construct every possible extension to the diagram you can think of, and write down a barrage of facts (unless you are really stuck), but an educated guess and a few sketches can work. Sometimes one can try using a special or extreme example to try to suggest a way (for instance, for the above problem, we could consider the case when $ABEF$ was square, or when $ABEF$ was degenerate, or when $|DC| = |DG| = 0$). And always keep the given data ($|DC| = |DG|$, $ABEF$ is a rectangle) and the objective ($|FC| \times |BC| = |AG| \times |DG|$, or some other formulation) in mind. Also try to steer your way towards unusual data or objectives. (In this case the strange-looking $DC = DG$). After all, one would presumably need all the data to deduce all the objectives, so each piece of data must be invoked in some way.

The key here was to recall a particular result in Euclidean geometry, in this case Theorem 4.2. With enough geometrical experience these things can come to mind after one has looked at every part of the problem and has 'grasped' the nature of the problem (these things also come to mind usually only after all other means have failed). Without such inspiration, one should stick to coordinate geometry or perhaps pseudo-coordinate geometry (drop perpendiculars from D to AB and AC, say, and use Pythagoras to represent $|DC|$ and $|DG|$—essentially coordinate geometry without the axes).

PROBLEM 4.4. Given three parallel lines, construct (with straightedge and compass) an equilateral triangle with each parallel line containing one of the vertices of the triangle.

At first glance this question looks simple and straightforward (good problems usually do). But as soon as one tries to draw a picture (try it, but draw the parallel lines first) one can see how tricky it really is to fit a triangle with as many requirements as an equilateral triangle. It is just too rigid. After experimenting with circles, 60° angles, and the like, we should see that something special is needed. Nevertheless, let us try to draw a picture as best we can (perhaps by drawing the equilateral triangle first, and then erasing it), and labelling everything:

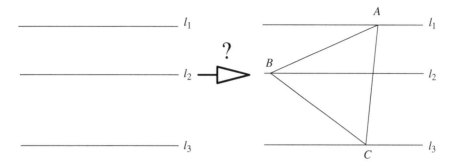

One obvious guess is to use coordinate geometry. Well, this is possible, but messy. You will end up using the quadratic formula to evaluate the positions of the points, and this is not the best (or most geometrical) way to do it. As usual, we save that as a last resort.

Well, the standard way to solve construction questions is to take one of the unknowns (a point, line, triangle, or maybe something else) and determine a locus or other easily constructible property.

But before we do that, let us just stare at the diagram and try to do what we can. One thing that can be seen is that an equilateral triangle, if there was one, could be slid along the parallel lines and still satisfy all the requirements. So, if the triangle was ABC, then the location of A is really arbitrary, as long as it is on line l_1. Of course, B and C will have to depend on the placing of A. So essentially we can place A wherever we like and not worry about missing anything, and then worry about B and C. A bit of thought now shows that the line l_1 is no longer relevant; it is only needed to constrain A, but once we pick A to be an arbitrary point on l_1, we do not need l_1 any more.

Now, with an anchor on A, the triangle is a bit more restricted. Perhaps this restriction could force B and C into a limited number of positions. We do not know yet.

The equilateral triangle has only two degrees of freedom now: its orientation and size. But it has two anchors: one point, B, must be on l_2, and the other point, C, must be on l_3. This theoretically should be enough to restrict the triangle, but with an object as complex as a triangle, it is hard to see where to go next. But what we can do is to shift the unknown into another unknown more easily evaluated. Currently the unknown is the equilateral triangle. What about something simpler? The simplest geometrical object is a point. So, we could work out B, for example, instead of the entire triangle. B has only one degree of freedom, as it is restricted to be on l_2. What is the anchor on B? The anchor is the fact that the equilateral triangle with base AB must have its third vertex (i.e. C) on l_3. This anchor is complicated and it still involves the equilateral triangle. Is there an easier way to represent

C in terms of A and B? Yes: C is the image of B after rotation of $60°$ through A (either clockwise or anti-clockwise). So the problem is reduced to this:

> Given a point A and two parallel lines l_2, l_3 not passing through A, find a point B on l_2 such that the rotation of B through A by $60°$ falls on l_3.

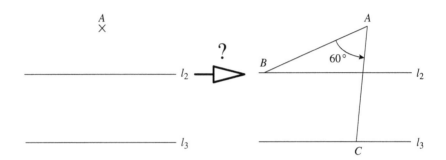

We now have only one unknown—the point B. So there are less degrees of freedom, and the question should be simpler. We want B to obey two properties:

(a) B lies on l_2.
(b) The rotation of B by $60°$ through A lies on l_3.

Condition (b) is not in a usable form, unless you invert it:

(b′) B is on the inverse rotation of l_3 by $60°$ through A.

that is, B is on $l_{3'}$, where $l_{3'}$ is the inverse rotation of l_3 through A by $60°$ (either clockwise or anti-clockwise). So our properties become:

(a) B is on l_2,
(b′) B is on $l_{3'}$,

or in other words B is the intersection point of l_2 and $l_{3'}$. And that is it! We have constructed B explicitly, so the triangle should easily follow.

Just for completeness, here is the entire construction:

> Choose any point A on l_1. Rotate l_3 about A by $60°$ (clockwise or anti-clockwise: there are two solutions of B for each given A) and let B be the intersection of that rotated line with l_2. Rotate B backwards by $60°$ to find C.

It can be noted that this construction also works if the lines were not parallel, so long as they are not at 60° angles to each other. So the parallelism was in fact a red herring!

With construction questions, the idea is to 'solve for' one's unknowns, in this case B, just like in algebra. We kept reformulating the data until it was in the form 'B is ...'. To give an algebraic analogy, suppose one wanted to solve for b and c given the following data:

- $b + 1$ is even
- $bc = 48$
- c is a power of 2.

Now if we solved for b in all three equations, and eliminate c, we get

- b equals an even number minus 1 (i.e. b is odd);
- b equals 48 divided by a power of 2 (i.e. $b = 48, 24, 12, 6, 3, 1.5, \ldots$).

And then by comparing the set of odd numbers to the set of all numbers formed by dividing 48 by a power of 2, we find that b is 3. It is often easier to solve questions with several variables by eliminating them one by one, and the same holds in constructional geometry.

EXERCISE 4.1. Let k and l be two circles that intersect in two points P and Q. Construct the line m through P, not containing Q, with the property that if m intersects k in B and P, and m intersects l in C and P, then $|PB| = |PC|$; see figure below. (Hint: solve for B.)

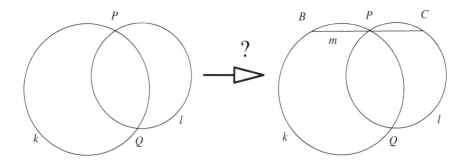

EXERCISE 4.2. We are given a circle and two points A and B inside the circle. If possible, construct a right-angled triangle inscribed in the circle such that one leg of the right-angled triangle contains A and the other leg contains B; see figure below. (Hint: solve for the right-angled vertex.)

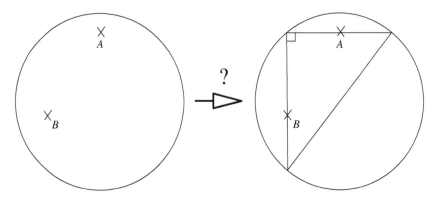

EXERCISE 4.3. (*) We are given four points A, B, C, and D. If possible, find a square so that each side contains one of the four points; see figure below. (Hints: unfortunately, it is very hard to solve for the square, and solving for a single vertex of the square (as one would do in the previous problems) is only a little better: the vertex can be confined to a fixed circle, but that is about it. One approach that does yield results is to solve for a diagonal of the square. A diagonal needs several anchors: the orientation, position, and endpoints. But the diagonal will determine the square uniquely, while a single vertex cannot do that easily. If you are really stuck, try drawing a nice big diagram with the square first and the points second, and then draw circles with AB, BC, CD, and DA as diameters, and also draw the diagonals. Use the circles to their fullest advantage: calculate angles, similar triangles, and so on. For a really big hint, look at the intersections of the diagonals and the circles. There is also another solution, where one solves for a particular side, by using rotations, reflections, and translations to twist one side to nearly match another. In short, a solution of similar style to the above.)

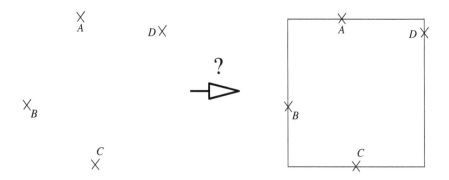

PROBLEM 4.5 (Taylor 1989, p. 10, Q4). A square is divided into five rectangles as shown below. The four outer rectangles R_1, R_2, R_3, R_4 all have the same area. Prove that the inner rectangle R_0 is a square.

Euclidean geometry

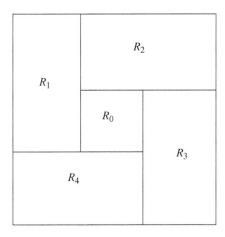

This is another of these 'unusual objective' questions. The fact that the outer four rectangles all have the same area does not seem to force the inner square to be equal, on first glance. At first you might think there is too much freedom in the data: after all, a rectangle with a fixed area could be long and thin, or short or fat. Why cannot we maneuvre one rectangle out of shape, and distort the inner rectangle? A quick try shows why this does not work: each rectangle is constrained by its neigbouring rectangles. In the picture, rectangle R_1, for example, is 'stuck in place' by rectangles R_2 and R_4. To change rectangle R_1 would involve changing rectangles R_2 and R_4, which would then both change rectangle R_3. But rectangle R_3 cannot satisfy the demands of both rectangles R_2 and R_4, unless they are demanding the same thing. In the following picture, rectangle R_3 can fit rectangle R_2, or rectangle R_4, but not both (remember R_3 also has to have the same area as R_2 and R_4). Light begins to dawn on how this question 'works': because of needs of equal area, as well as the difficulties of 'flush fitting', the only possibile way this can work is if the inner rectangle is a square. It should be impossible to move out of this symmetric swastika formation; the figure below gives an example of what can go wrong.

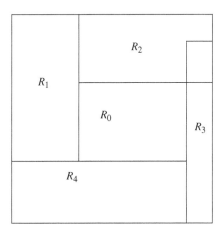

To progress any further, we need notation: more specifically we need to express all the various sizes and dimensions of the geometrical objects in terms of a few variables. From our discussion of 'wriggling' the formation it is apparent that one rectangle, for example rectangle R_1, will determine the positions of all the other rectangles; R_1 will force R_2 and R_4 into fixed positions, which will in turn fix rectangle R_3, if possible. So there is an algebraic approach: assume that the dimensions of rectangle R_1 are, say, $a \times b$, that the big square has side length 1, and determine the dimensions of all other rectangles, and in particular R_0. This is the sledge-hammer approach: we will end up with two equations concerning rectangle R_3 (or perhaps R_1, R_2, or R_4 if we interpret the equations differently) and then we can solve for a relation between a and b (for not just any dimensions of R_1 will work: in fact we have to prove that the only formations of R_1 allowed are those that produce a square in the middle). The following diagram summarizes the situation.

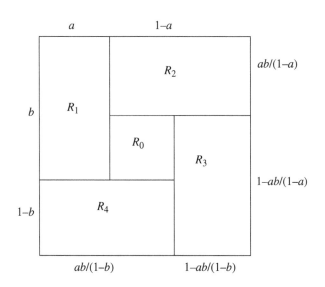

In order for R_3 to have the right area, we have $(1 - ab/(1 - a)) \times (1 - ab/(1 - b)) = ab$, and one can use this to solve for a, b and then determine that R_0 is a square. This method works, but is a bit messy algebraically, so let us try for a simpler, more intuitive, and less coordinate-based approach (which is really all that approach is).

We want to prove that the only way all our conditions are satisfied is when R_0 is a square. But that is a bit hard to prove. We have already shown that we can put everything in terms of, say, rectangle R_1. In this sense, rectangle R_1 can be called the main figure: the ones that all the other constructions depend on. Once we have this reference point, we can concentrate on one rectangle alone. So, instead of trying to prove something about rectangle

R_0, which does not become a 'main figure' as easily as other rectangles, we can prove something about rectangle R_1, which should be easier to prove.

The above picture seems to suggest that $a + b$ should equal 1. Indeed, if $a + b$ equalled 1, then R_2 must have a horizontal length of $1 - a = b$, by equal area must have had vertical length of a, so that rectangle R_3 must have had a vertical length of $1 - a = b$, and so on. It fits very neatly into the above-mentioned 'swastika', and one sees that R_0 is a square with sidelength $b - a$. So we have isolated an intermediate goal, of showing that $a + b = 1$. Heuristically, we hope this goal is easier to achieve because we can put everything in terms of a and b, while it is not that easy to put everything in terms of rectangle R_0.

To summarize, we have shown the second implication in the chain

$$\boxed{R_1,\ldots,R_4 \text{ equal area}} \implies \boxed{a+b=1} \implies \boxed{R_0 \text{ is a square}}$$

and it now remains to prove the first implication.

We can see by the coordinate geometry approach that, while the given data is easily reducible to formula, the formula is not easily reduced to objective. While equal areas may seem to be a very nice and simple thing to work with, they are actually more of a hindrance in this question, because you just have a bunch of equal products whose terms are related by additive equations. But we can work backwards: we can try to prove that

$$\boxed{a+b \neq 1} \implies \boxed{R_1,\ldots,R_4 \text{ do not have equal area}}$$

or we could try a proof by contradiction:

$$\boxed{a+b \neq 1} \,\&\, \boxed{R_1,\ldots,R_4 \text{ have equal area}} \implies \boxed{\text{Contradiction}}$$

Notice that with a proof by contradiction, one has more data to begin with, but the end result is very open-ended and indefinite. This strategy fits well with our earlier, qualitative approach: it would be impossible to wriggle from the symmetric solution because all the rectangles get unbalanced. So let us focus a bit more on the proof by contradiction method.

So suppose that $a + b$ is too large: that it is greater than 1, but the rectangles somehow manage to have the same area. Then we have to prove a contradiction. Well, what we have is a rather big rectangle R_1. What does that do? it forces rectangle R_2, say, to be narrow. In fact the horizontal length of R_2 is $1 - a$, and is smaller than b. Therefore, R_2 is narrower than R_1. Because of the equal areas restriction, R_2 has to be longer than R_1 in the vertical direction. So R_2 is more stretched out than R_1. But now look at rectangle R_3: by similar logic it must be more 'stretched out' than rectangle R_2. And applying the same reasoning again, rectangle R_4 must be stretched further out than rectangle R_3. And one last time: rectangle R_1 must be thinner and longer than rectangle R_4. But this means that rectangle R_1 is longer and thinner than itself, which is absurd. And here we have our contradiction. A similar condition occurs when $a + b$ is less than 1: the rectangles get fatter and shorter, and eventually one can show rectangle R_1 has to be more squashed than itself, which is the contradiction.

This question is a good example of how a picture is worth a thousand equations. Also, keep in mind that sometimes inequalities are easier and more efficient to use than equalities.

EXERCISE 4.4. Find all positive reals x, y, z and all positive integers p, q, r such that

$$x^p + y^q = y^r + z^p = z^q + x^r.$$

(Hint: This question has no geometry in it, but it is still similar to Problem 4.5.)

PROBLEM 4.6 (AMOC Correspondence Problem, 1986–1987, Set One, Q1). Let $ABCD$ be a square, and let k be the circle with centre B passing through A, and let l be the semicircle inside the square with diameter AB. Let E be a point on l and let the extension of BE meet circle k at F. Prove that $\angle DAF = \angle EAF$.

As always, we begin by drawing a picture:

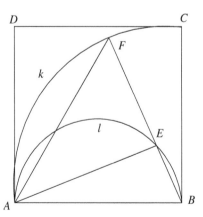

Now we have to equate two angles. Judging by the lack of side lengths and so on, it seems we can tackle the question entirely by angles. After all, circles are always very friendly towards angles. But these particular angles $\angle DAF$, $\angle EAF$ seem a bit remote. We will need to write these obscure angles in terms of more 'friendly' angles, so that we can then relate the two angles to each other.

Let us take $\angle DAF$ for a start. The angle $\angle DAF$ does not connect to any triangles, but it does connect to circle k. The odd little theorem (Euclid III, 32) that the angle a chord subtends in a circle is the same as the angle the chord subtends in the tangent can be used here: we can say that $\angle DAF = \angle APF$, where P is any point on k which is on the arc of AF containing D. For example, we could say $\angle DAF = \angle ACF$. The $\angle ACF$, though, is almost as boring as $\angle DAF$. But it is an angle subtended on a circle. This means it is half the angle subtended by the centre: that is, $\angle ACF = \frac{1}{2}\angle ABF$. The angle $\angle ABF$ seems to be a more 'mainstream' angle, connected to several triangles and circles, so this is a fairly satisfying result: $\angle DAF = \frac{1}{2}\angle ABF$.

Now we can tackle $\angle EAF$. This angle is even more ugly than $\angle DAF$; it is not directly attached to anything else. But it shares its vertex with other, nicer angles like $\angle DAB$, $\angle EAB$, and the like, so we can represent $\angle EAF$ in terms of friendlier angles, for instance as

$$\angle EAF = \angle BAF - \angle BAE$$

or perhaps

$$\angle EAF = \angle DAB - \angle DAF - \angle BAE.$$

The first equation leaves us with one rather nice angle $\angle BAE$ and one slightly worse angle $\angle BAF$. The second formulation, however, has several advantages: $\angle DAB$ is 90°, and we have already worked out $\angle DAF$. So we get

$$\angle EAF = 90° - \frac{1}{2}\angle ABF - \angle BAE.$$

But $\angle BAE$ and $\angle ABF$ are in the same triangle ABE. Since we have written both $\angle DAF$ and $\angle EAF$ are now written in terms of angles from ABE, it is clearly time to start focusing on this triangle.

Well, ABE is inscribed in a semicircle. This should remind one of Thales' theorem (Theorem 4.1), which tells us that $\angle BEA = 90°$. This ties together the angles $\angle ABF$ and $\angle BAE$ now, because the sum of the angles of a triangle is 180°. To be precise, we have $\angle ABF + \angle BAE + \angle BEA = 180°$, hence $\angle BAE = 90° - \angle ABF$. Now we can plug this back into our expression for $\angle EAF$, and get

$$\angle EAF = 90° - \angle ABF/2 - \angle BAE = 90° - \frac{1}{2}\angle ABF$$
$$- (90° - \angle ABF) = \frac{1}{2}\angle ABF.$$

But this is just the same expression we have for $\angle DAF$. So we have proved $\angle EAF = \angle DAF$. Of course, we will want to tidy this up when presenting our proof: we would probably do some long chain of equations like so:

$$\angle DAF = \vdots$$
$$= \vdots$$
$$= \vdots$$
$$= \angle EAF.$$

But when we are looking for a solution, we do not have to be so formal. Working out $\angle DAF$ and $\angle EAF$ and hoping that they meet somewhere in between is not all that foolish, if you know what you are doing. As long as one always tries to simplify and connect, chances are that the solution will soon fall into place. (Assuming, of course, that there is one—and most problems are not trying to pull your leg.)

5 Analytic geometry

> The geometrical mind is not so closely bound to geometry that it cannot be drawn aside and transferred to other departments of knowledge. A work of morality, politics, criticism, perhaps even eloquence will be more elegant, other things being equal, if it is shaped by the hand of Geometry.
>
> *Bernard le Bovier de Fontenelle*, 1729, 'Preface sur l'Utilite des Mathematiques et la Physique', translated by F. Cajori

This chapter consists of problems involving geometrical concepts and objects, but whose solution requires ideas from other branches of mathematics: algebra, inequalities, induction, and so on. One nice trick that sometimes works very well is to rewrite the geometry problem in terms of vectors in order to use the laws of vector arithmetic. Here is an example of this.

PROBLEM 5.1 (Australian Mathematics Competition 1987, p. 14).
A regular polygon with n vertices is inscribed in a circle of radius 1. Let L be the set of all possible distinct lengths of all line segments joining the vertices of the polygon. What is the sum of the squares of the elements of L?

First of all, let us give 'sum of squares of the elements of L' a shorter name, such as 'X'; our job is to compute X. This is what might be called a 'feasible' problem: It is not a 'prove this' or 'is this' problem, but an evaluation of a number that could be found, by say a straightforward application of trigonometry and perhaps Pythagoras' theorem. For example when $n = 4$, we have a square inscribed in the unit circle. The possible lengths are the side length $\sqrt{2}$ and the diagonal length 2, so $X = \sqrt{2}^2 + 2^2 = 6$. Similarly, when $n = 3$, the only length we have is the side length, which is $\sqrt{3}$, so in this case $X = \sqrt{3}^2 = 3$. $n = 5$ is not as simple to evaluate unless you know a bunch of sines and cosines, so let us skip it and try $n = 6$ instead. The side length is 1, the small diagonal is $\sqrt{3}$, and the long diagonal (the diameter) is 2. Hence in this case $X = 1^2 + \sqrt{3}^2 + 2^2 = 8$. Finally, we have the rather degenerate case $n = 2$, in which case the 'polygon' is just a diameter, and

$X = 2^2 = 4$. So we have some special cases worked out:

n	X
2?	4?
3	3
4	6
6	8

where we have marked the $n = 2$ case with a question mark as it is a little dodgy to talk about a two-sided polygon.

This small table does not give us too many clues about what the general answer should be. The first thing is to draw a picture. It probably makes good sense to label the vertices. For a fixed value of n (say $n = 5$ or $n = 6$) one could label the vertices A, B, C, etc., but for the general n case it is probably going to be more convenient to label the vertices $A_1, A_2, A_3, \ldots, A_n$, like so:

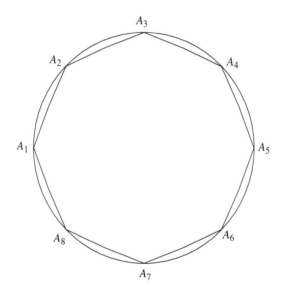

Now we can make some initial observations and guesses:

(a) It could make a difference if n is odd or even. If n is even, we have this long diagonal to contend with. In fact when n is even we have $n/2$ diagonals, and when n is odd we have $(n-1)/2$ diagonals.

(b) The answer could always be an integer. This is not a strong conjecture at present, because we are dealing with the very special square, hexagon, and equilateral triangle which have square-root type measurements. However, it gives us a bit of hope that the general solution would not be too nasty.

(c) We are summing the squares of lengths, not lengths themselves. This immediately puts us out of the realm of pure geometry and into analytic geometry: it suggests vectors, or coordinate geometry, or perhaps complex numbers (these are essentially all the same approaches anyway). Coordinate geometry will give a slow-but-steady solution involving trigonometric sums, but vector geometry and complex numbers both look promising (vector geometry can use the dot product and complex numbers can use the complex exponential).

(d) It is almost impossible to try to deal with the problem directly, because we are not summing the squares of the lengths of all diagonals, only all diagonals of distinct length. But we can soon rephrase the question into a form which is more reducible to an equation. (Equations are solid mathematics. Not as inspirational as pictures and ideas, but the easiest to manipulate. In general, one would always express the objective as an equation of some sort. Some possible exceptions, though, are in combinatorics or graph theory.) However, if you restrict yourself to the diagonals emanating from a single point of the polygon, these diagonals will cover all the lengths we need.

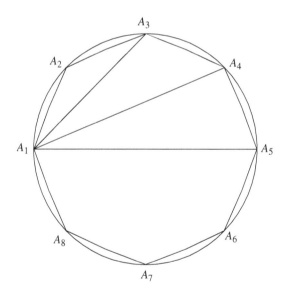

For example, in this figure, with n even, we have the four different lengths of diagonal. If you now just restrict yourself to the upper semicircle, then each length of diagonal is encountered exactly once; the lengths $|A_1A_2|$, $|A_1A_3|$, $|A_1A_4|$, and $|A_1A_5|$ will cover the lengths we want. In other words, the answer can be given as an expression: $|A_1A_2|^2 + |A_1A_3|^2 + |A_1A_4|^2 + |A_1A_5|^2$. More generally, we are trying to compute $|A_1A_2|^2 + \cdots + |A_1A_m|^2$, where $m = (n/2) + 1$ (if n is even) or $m = (n+1/2)$ (if n is odd). So we have a more explicit way of stating the problem:

> Let a regular polygon with n vertices $A_1, A_2, \ldots, A_n$ be inscribed in a circle of radius 1. Let $m = (n/2) + 1$ if n is even, or $m = (n+1)/2$ if n is odd. Compute the quantity $X = |A_1 A_2|^2 + \cdots + |A_1 A_m|^2$.

It is a bit inconvenient that the sum $|A_1 A_2|^2 + \cdots + |A_1 A_m|^2$ stops at A_m, rather than the more natural A_n. But (as with Problem 2.6) we can 'double up'; from symmetry we have $|A_1 A_i| = |A_1 A_{n+2-i}|$, and so

$$X = \frac{1}{2}(|A_1 A_2|^2 + |A_1 A_3|^2 + \cdots + |A_1 A_m|^2 + |A_1 A_n|^2 + |A_1 A_{n-1}|^2 + \cdots + |A_{n+2-m}|^2).$$

Note that when n is even, we will have counted the diagonal $|A_1 A_{n/2+1}|^2 = 4$ twice. We can tidy this up, and add the quantity $|A_1 A_1|^2$ (since it is zero) for sake of symmetry, to get

$$X = \frac{1}{2}(|A_1 A_1|^2 + |A_1 A_2|^2 + \cdots + |A_1 A_n|^2) \qquad (16)$$

when n is odd, and

$$X = \frac{1}{2}(|A_1 A_1|^2 + |A_1 A_2|^2 + \cdots + |A_1 A_n|^2) + 2 \qquad (17)$$

when n is even (the 2 comes from the extra diagonal term $|A_1 A_{n/2+1}|^2 = 4$, multiplied by $\frac{1}{2}$). It is now natural to introduce the quantity

$$Y = |A_1 A_1|^2 + |A_1 A_2|^2 + \cdots + |A_1 A_n|^2 \qquad (18)$$

and try to compute Y instead of X. The advantages of doing this are:

- Once one knows Y, the equations (16) and (17) will immediately give us X.
- Y has a prettier form than X, and thus is hopefully easier to compute.
- When computing Y, we do not have to split up into cases when n is even or odd, which may save us some work.

We can revisit our previous table of the small cases $n = 3, 4, 6$ and compute Y in these cases (for instance, by using (16) and (17)):

n	X	Y
2?	4?	4?
3	3	6
4	6	8
6	8	12

From this table one can now conjecture that $Y = 2n$; from (16) and (17), this would mean that $X = n$ when n is odd and $X = n + 2$ when n is even. This is probably going to be the right answer, but we still have to prove it.

Now it is time to use vector geometry, as it gives us a number of useful tools to manipulate expressions such as (18). Because the square of a length of a vector v is simply the dot product $v \cdot v$ of v with itself, we can write Y as

$$Y = (A_1 - A_1) \cdot (A_1 - A_1) + (A_1 - A_2) \cdot (A_1 - A_2) + \cdots$$
$$+ (A_1 - A_n) \cdot (A_1 - A_n),$$

where we now think of $A_1, \ldots, A_n$ as vectors rather than points. We can choose the origin of our coordinate system to be wherever we want, but the most logical choice would be to put the origin at the centre of the circle. (The second best choice would be to make A_1 the origin.) An immediate advantage of putting the origin at the centre of the circle is that all the vectors $A_1, \ldots, A_n$ have length 1, thus $A_1 \cdot A_1 = A_2 \cdot A_2 = \cdots = A_n \cdot A_n = 1$. In particular, we can use vector arithmetic to obtain

$$(A_1 - A_i) \cdot (A_1 - A_i) = A_1 \cdot A_1 - 2A_1 \cdot A_i + A_i \cdot A_i = 2 - 2A_1 \cdot A_i$$

and so we can expand out Y as

$$Y = (2 - 2A_1 \cdot A_1) + (2 - 2A_1 \cdot A_2) + \cdots + (2 - 2A_1 \cdot A_n).$$

We can gather some terms and simplify to

$$Y = 2n - 2A_1 \cdot (A_1 + A_2 + \cdots + A_n).$$

Now we had guessed that $Y = 2n$. We can now achieve this guess if we can show that the vector sum $A_1 + A_2 + \cdots + A_n$ is zero. But this is clear from symmetry (the vectors 'pull' in all directions with equal strength, so the net result has to be 0. Alternatively, you could say that the centroid of the regular polygon is equal to its centre. Always look for ways to exploit symmetry.) So $Y = 2n$, and so we indeed have $X = n$ for n odd and $X = n + 2$ for n even.

One may not be happy with the hand-waving symmetry argument that $A_1 + \cdots + A_n = 0$. One could use trigonometry or complex numbers to give a more concrete proof, but here is a more explicit symmetry argument that may be more satisfying: write $v = A_1 + \cdots + A_n$. Now rotate everything around the origin by $360°/n$. This shifts all the vertices $A_1, \ldots, A_n$ by one, but that does not change the sum $v = A_1 + \cdots + A_n$. In other words, when we rotate v by $360°/n$ around the origin we get back v again. The only way this can be true is of $v = 0$, and so $A_1 + \cdots + A_n = 0$ as desired.

One can interpret the above argument in a more physical manner; indeed, the sum of squares Y is basically the moment of inertia about A, and then one can use Steiner's theorem of parallel axes (Borchardt, 1961, p. 370) to move the rotation point to the centroid.

EXERCISE 5.1 (**). Prove that the area of a unit cube's projection on any plane equals the length of the cube's projection on the perpendicular of this plane. (Hint: There is a neat vector solution, but it involves a good track keeping of cross products and the like. First of all, select a good coordinate system, then choose the most friendly vectors you can find. Then write down the objective and manipulate it, using cross products, dot products, and the huge number of pairs of perpendicular vectors to your advantage. Also exploit the fact that many vectors v have unit lengths (so that $v \cdot v = 1$). Eventually, once you have solved it, you can rewrite the proof and then see how a resolution by vectors can be abstractly neat.)

PROBLEM 5.2. (*) A rectangle is partitioned into several smaller rectangles. Each of the smaller rectangles has at least one side of integer length. Prove that the big rectangle has at least one side of integer length.

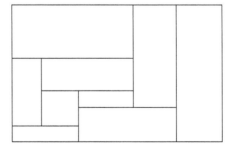

This is a pleasant-looking problem, so presumably there should be a pleasant solution. But the conclusion is a bit odd: if all the small rectangles have one integer side (or maybe more), why should the big rectangle have one? If we were not playing with rectangles, but just line segments, the thing would be easy: the big segment is composed of little segments, each of integer length, so the length of the big segment is a sum of integers which is an integer, of course. This one-dimensional case does not immediately offer any help to the two-dimensional case clearly, except we get the following clue: we have to use the fact that the sum of integers is an integer. One way we can use this fact immediately is to get some convenient notation: an 'integer side' means a side of integer length.

But the question also has traces of topology, combinatorics, and worse, mainly because of the word 'partition'. It is a bit too general. To get a handle on this question, let us try the simplest (but non-trivial) partitioning as follows:

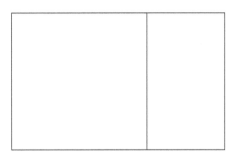

We have two sub-rectangles, and we know that they each have at least one integer side. But those sides could be horizontal or vertical: we do not know which. Suppose the sub-rectangle on the left had a vertical integer side. But its vertical side length is the same as the big rectangle's vertical side length, so we have proved de facto that the big rectangle has an integer side. So we can assume that the left-hand rectangle has a horizontal integer side instead.

But we can argue by similar reasoning that the right-hand rectangle has a horizontal integer side, and so the big rectangle has a horizontal integer side, because it is the sum of two rectangles with horizontal integer sides. So we have proved this question for the special case of the two-rectangle partition. But how did it work? (Examples are really only useful when they give some insight into how the general problem works.) Scanning the above proof, we observe two key ingredients:

(a) We have to split into cases, because each sub-rectangle can have a vertical integer side or a horizontal integer side.
(b) The only way we can prove the big rectangle to have a vertical integer side, say, is if a 'chain' of smaller rectangles have a vertical integer side, and if the smaller rectangles somehow 'add up' to the big rectangle. Here

is an example, where the shaded rectangles have a horizontal integer side, so the big rectangle has to have a horizontal integer side as well:

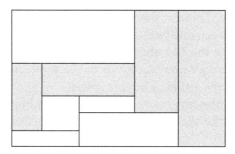

So with these vague aims in mind, we can formulate this vague strategy:

> Find either a chain of horizontal-integer-side-rectangles or a chain of vertical-integer-side-rectangles, that somehow 'add up' into a horizontal integer side or a vertical integer side for the big rectangle.

But we have to find such a chain for every possible partition. Partitions are very ugly things. And each rectangle has a choice of a horizontal integer side, or a vertical integer side. Some rectangles may have both. Now how could we possibly find a system that will work for all these possibilities?

How do these chains work anyway? If several small rectangles have a horizontal integer side, the big rectangle will have a horizontal integer side if the small rectangles 'link' from one end of the rectangle to another, as in the above figure, because then the length of the big edge is just the sum of the lengths of the smaller edges. (In other words, if you stack some blocks on top of each other, the total height of the structure is the sum of the height of the blocks.)

Part of the problem in finding these chains is that we do not know which rectangles have horizontal integer sides, and which rectangles have vertical integer sides. To visualize the possibilities, imagine that any rectangle with a horizontal integer edge is coloured green, and any rectangle with a vertical integer edge can be coloured red. (Rectangles with both horizontal and vertical integer edges are a bonus: they can be assigned either colour.) Now each small rectangle is coloured either green or red. And now we have to find a green chain connecting the two vertical edges or a red chain connecting the two horizontal edges.

No direct proof seems to be available, so let us try proof by contradiction. Suppose that the two vertical edges are not connected by green rectangles. Why cannot they be connected? Because there are not enough green rectangles: the red rectangles must have blocked the greens. But the only way to block the green rectangles from reaching the vertical edge is by a solid barrier of red rectangles. But a solid barrier of red rectangles must join the two horizontal edges. So either the greens span the vertical edges or the reds span the horizontal edges. (Anyone who is familiar with the game Hex may see comparisons here.)

(Incidentally, while the essence of the above paragraph is a very intuitive statement, actually proving it formally and topologically requires some work. Briefly: the set of all green areas can be divided into connected subsets. Assuming that none of these subsets straddle both vertical edges, consider the union of the left vertical edge together with all green connected subsets that touch on the left vertical edge. Then a small strip touching the boundary of this set on the outside will be coloured red, and this red strip defines a red set of rectangles which will straddle the two horizontal edges.)

Now there is a small matter of checking that chains of green rectangles spanning the vertical edges does ensure that the big rectangle has a horizontal integer edge. The only real problems are superfluous rectangles in a chain, which are easily ditched; rectangles that touch only on a corner, which is also not a problem; and backwards-going chains, which are easily dealt with too (we are subtracting integers instead of adding them on, but the grand total will always be an integer).

PROBLEM 5.3 (Taylor 1989, p. 8). On a plane we have a finite collection of points, no three of which are collinear. Some points are joined to others by line segments, but each point has at most one line segment attached to it. Now we perform the following procedure: We take two intersecting line segments, say AB and CD, and remove them and replace them with AC and BD. Is it possible to perform this procedure indefinitely?

First of all, one should check that this procedure does not lead to any degenerate or ambiguous situations, in particular we do not want to create segments of zero length, or have two segments coincide with each other. This is why we have the condition 'each point has at most one line segment attached to it'. Anyway, it is easy to verify but should be done nevertheless (it could be a tricky question!).

After trying a few examples it seems that the assertion is plausible. The line segments seem to all hide away in the outer edges after a while and no longer intersect. That is easy to say in English, but how do we say that in Mathematics?

We have to say that somehow the 'outeredginess' of the system must increase each time we perform the procedure. But this cannot happen indefinitely because there are only a finite number of configurations the system can reach. Eventually the 'outeredginess' should reach at a maximum and the procedure will stop. (That is, things stop when they cannot go any further.)

So now we have to do the following things:

(a) Find some characteristic of the system that can be represented as a number. It could be number of intersections, number of line segments, perhaps a sum of carefully chosen point scores (like a dart board). It must reflect 'outeredginess'; that is, it should be higher when all the edges are scattered on the rim.
(b) This characteristic must increase (or perhaps remain steady, but this is a lot weaker) every time our procedure is used.

[For example, anyone familiar with the game *Sprouts* may see that every time a move is made (joining two points and placing a third) the number of available exits (an exit is an unused edge from a point: each point starts with three exits) goes down by one (two are used up by the line and one is created by the new point). This shows that the game cannot last forever, because one will run out of exits.]

Now we have to find a characteristic that satisfies (a) and (b). There is no unique solution: several kinds of characteristics would satisfy (a) and (b). But we only need one. The best technique is just guessing something simple and hope that it works.

Let us try the really simple first. How about 'number of points'? That never changes, so that does not help. 'Number of lines' fail for the same reason. 'Number of intersections' looks promising, but the number of intersections does not always decrease with each application (although it should decrease in the long run), as can be seen here:

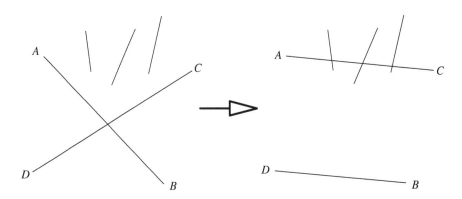

One point of intersection has become three.

After all, what decreases when one changes two intersecting lines into two non-intersecting lines? Somehow the lines become more separate. At this point one may try something like 'sum of distances betweeen line segments', but this is not easy to do. But in a similar spirit, we can eventually happen across 'sum of lengths of line segments'; not only do the line segments become more separate, they become *shorter* as well. (The triangle inequality—the sum of two sides of a triangle is always longer than the third—shows that quite nicely.) This means that the total length of all edges must shrink after each operation, and hence the operation cannot cycle or go on forever (as there are only a finite number of possibilities for the edges connecting the vertices, which are fixed), and so the problem is solved.

Because we are changing two line segments each time, any characteristic under consideration should be in terms of individual line segments, rather than intersections or other properties. Now there are really only three properties of individual line segments: length, position, and orientation. Position and orientation type characteristics fail to give nice results because those properties cannot really decrease or increase: for example, it is unlikely that, after each operation, the total orientation (whatever that is) goes clockwise. If that were so, why clockwise and not counter-clockwise? There is no real distinction between clockwise and counter-clockwise, while there is a definite difference between longer and shorter. With this in mind, one almost is forced to use 'total length' ideas.

PROBLEM 5.4 (Taylor 1989, p. 34, Q2). In the centre of a square swimming pool is a boy, while his teacher (who cannot swim) is at one corner of the pool. The teacher can run three times faster than the boy can swim, but the boy can run faster than the teacher can. Can the boy escape from the teacher? (Assume both persons are infinitely manoeuvrable.)

Let us first draw a picture and label some points:

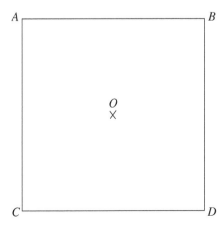

So the boy starts at O, and the teacher starts at one of the corners—say A. We can also choose the units of length so that the sides of the pool are one unit long.

Now, to solve this question, we should first decide for ourselves what the answer is going to be. (You cannot really search for a solution unless you know what you are searching for.) Circumstantial evidence is a bit uncertain: if the boy could escape, then there would have to be a winning strategy for the boy. Otherwise, there would be a winning strategy for the teacher, who could somehow always maneuvre to intercept him no matter what. The latter possibility is a bit grim mathematically: we have to find a strategy that would defeat every possible move the boy could make—and there are many, many options for the boy (who can move in two dimensions, while the teacher is effectively limited to one). But the first possibility is easier without becoming too trial-and-errorish: we would have to intelligently guess a strategy, and then prove that it works. Surely it is easier to prove that one strategy works, than to prove that all strategies do not work. So, let us assume the boy can escape. It looks the easier of the two options: always tackle the soft options first—you may then be able to dodge a load of hard work later on. (This is not laziness, but practicality. As long as the job is done well, the easy way is of course better than the hard way.)

The boy can run faster than the teacher, so this means that once he is on land and unintercepted, he can escape. So his first goal is to get out of the pool. Now that we have shown that, the boy's running speed is now irrelevant.

Before we start guessing strategies, let us use some common sense to eliminate some bad strategies and isolate some promising tactics. First of all, the boy should use his maximum movement rate: even if some obscure advantage is gained by slowing down, the teacher can simply slow down to match. A similar argument shows that stopping is useless: the teacher can wait until the boy moves again. (Stalemate is not a victory from the boy's point of view.) Second, we may assume that the teacher is no pushover, and will stick to the edge of the pool (why go away from the edge? It would only slow the teacher down). Third, since the boy is trying to reach the edge of the pool quickly (or, at least quicker than the teacher could), so presumably straight lines, being the shortest (and hence the fastest) way to move, are part of the answer—although sudden turns and twists could conceivably be used to the boy's advantage. Finally the strategy should not be wholly predetermined, but depend in part on the actions of the teacher: after all, if the teacher knows that you will wriggle around for a while until you reach corner B, for example, then the teacher could simply run to corner B and wait for the boy to arrive—if he follows his rather foolishly pre-determined plan.

To summarize: The boy's best strategy would involve straight line bursts at top speed. Also, the strategy could be flexible to account for the actions of the teacher.

With these general guidelines in mind, we can try some strategies. Obviously the boy will have to move away from the teacher: heading straight for A is not a clever move. Now the intuitive response then is to head straight for C, being the furthest away from A. The boy has to swim for $\sqrt{2}/2 \approx 0.707$ units of length, while the teacher has to run from $A \to B \to C$ or $A \to D \to C$, crossing two lengths of the pool to reach the boy's exit. The teacher, being three times faster, will reach C when the boy has only travelled $2/3 \approx 0.667$ units of length. So this approach does not work: the teacher arrives first.

Sneakier manoeuvres are needed than simply running away. After all, people trying to leave a pool head for the edges, not for the corners. Let us try, for example, to head to the midpoint M of B and C. Then the boy need only swim $1/2 = 0.5$ units of length. But the teacher does not have to run as much either: from $A \to B \to M$ is a distance of 1.5 units. The teacher, being three times as fast, will just barely catch the kid as he is climbing out of the pool.

The teacher almost lets the boy escape when he heads for the edge: if the teacher was only a fraction slower, then the boy would escape. This suggests that either

- the teacher's speed is the barest minimum needed to stop the boy;
- the teacher's speed is the barest maximum which allows the boy to escape.

This complicates matters a bit, because the teacher's speed seems to be on a knife-edge here. If the teacher was marginally slower, the boy would automatically escape if he simply headed for the edge. If the teacher was significantly faster, it is possible that the teacher could stalk the boy, moving clockwise when the boy moves clockwise, and so on. Common sense fails to reach a verdict; we have to do some computation.

If the boy makes for an edge, the teacher has to run continously just to keep even. In other words, the boy is forcing the teacher's moves just by threatening to move in one direction. Being able to control, to some degree, the enemy moves can be a potent tool. Can we use it?

Suppose the boy is heading towards the midpoint M of BC at full speed. The teacher cannot afford to do anything other than run towards B and then towards M. If the teacher changes direction, or does anything else, the boy can keep going and reach the edge before the teacher. But the boy does not have to go all the way to the edge—the threat is enough to send the teacher running. The upshot of this is that the boy can force a situation like this, if he swims at full speed and reaches some intermediate point X:

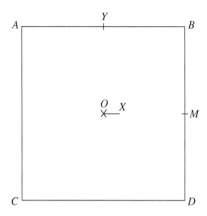

The teacher must be at point Y (Y being such that $|AY| = 3|OX|$): the teacher is not fast enough to move past Y by the time the boy has reached X, and if he has not reached Y yet, then the boy can continue all the way to M and escape the teacher. So the teacher has to go to Y.

Now what we have is that the boy is at X, the teacher being forced to be at Y. Now is there any real need to continue all the way towards M? The threat of being able to go to M was enough to keep the teacher where he is now, but threat and actuality are different things. Now that the teacher is stranded in the AB edge, why does not the boy make a dash for the opposite CD edge? It only takes half a length to reach the edge, and unlike the first time we considered the boy dashing for the edge, the teacher is badly placed. In fact, if X is a quarter-length or less from M, then it is easily shown that the teacher is just too far away to catch the boy. So, the boy escapes quite easily.

EXERCISE 5.2 (*). Supposer the teacher can run six times as fast as the boy can swim. Now show that the boy cannot escape. (Hint: Draw an imaginary square of sidelength 1/6 unit centred at O. Once the boy leaves that square, the teacher gains the upper hand.)

EXERCISE 5.3 (**). Suppose the pool was circular instead of square. Now it is clear that the boy can escape (just head for the point opposite the teacher). But what if the teacher was faster? More precisely, what is the minimum speed of the teacher needed to catch the boy? This needs a fair whack of creativity (or knowledge of calculus of variations) for finding a lower bound (i.e. designing escape strategies for the boy) and for calculating an upper bound (which needs a perfect set of moves for the teacher). (One could also ask the same problem for the square pool, but that is even trickier than the circular problem.)

6 Sundry examples

Mathematics is sometimes thought of as a great entity, like a tree, branching off into several large chunks of mathematics, which themselves branch off into specialized fields, until you reach the very ends of the tree, where you find the blossoms and the fruit.

But it is not easy to classify all of mathematics into such neat compartments: there are always fuzzy regions in between branches and also extra bits outside all the classical branches.

The following questions are not quite game theory, not quite combinatorics, and not quite linear programming. They are just a bit of fun.

> PROBLEM 6.1 (Taylor 1989, p. 25, Q5). Suppose on a certain island there are 13 grey, 15 brown, and 17 crimson chameleons. If two chameleons of different colour meet, they both change to the third colour (e.g. a brown and crimson pair would both change to grey). This is the only time they change colour. Is it possible for all chameleons to eventually be the same colour?

The question is a bit open-ended, with the 'eventually' in the question. This means that we have to decide whether the set of all possible chameleon colour combinations includes a state where all chameleons are a single colour.

Heuristically, we should first try the possibility that the answer is NO. If the answer is YES, then there should be a specific procedure to reach our objective. That sounds more computational than mathematical, and given that this question occurred in a mathematics tournament there is good grounds that this is not the right answer. So let us try to prove NO.

To prove this, it is probably a good idea to know which systems the procedure can reach, and which ones it presumably cannot. Once we have found a pattern, we will have something definite to prove. As we saw in the previous chapters, to solve a problem in mathematics, you usually have to guess some intermediate result, which implies the conclusion but is not logically equivalent to it. Although from a logical point of view this leaves you with a problem that may be harder to prove, pragmatically it should provide an objective that is nearer to our data and would concentrate our efforts in a more definite direction. Generalizing the conclusion tends to remove superfluous information as well, a further bonus.

For a simple example, suppose on a chessboard we have one bishop on the corner (a bishop moves diagonally), and we have to show that it can never move on an adjacent corner (i.e. either of the two corners not opposite it). Instead of proving that, we could have proved the more general 'the bishop must stay on the same colour square'. (The chessboard is checkered.) Logically, there is more to prove; but now it is very easy to see how to proceed (each move of the bishop keeps it in the same colour square; therefore, no number of moves will ever leave that colour of square).

Anyway, let us have some decent notation first (i.e. numbers and equations). At any given time, the only data which is important is the number of chameleons that are grey, the number which are brown, and the number which are crimson. (The set-up of the problem does not permit the chameleons to take on any additional colours.) We can represent this information efficiently by means of a three-dimensional vector; thus the initial state of the chameleons is $(13, 15, 17)$, and the problem asks if we can reach $(45, 0, 0)$, $(0, 45, 0)$, or $(0, 0, 45)$ by our operation of changing colour. The operation of changing colour consists of subtracting 1 from two of the coordinates and adding 2 to the third one. So here we have a vector formulation, which is actually one way of attacking the problem.

(For a brief sketch of the proof here, let $a = (-1, -1, 2)$, $b = (-1, 2, -1)$, and $c = (2, -1, -1)$. Then the meeting of two chameleons is represented by adding one of the vectors a, b, or c, to the current 'state vector'. Therefore, any position that the system can reach must have a position vector of the form $(13, 15, 17) + la + mb + nc$, where l, m, and n are integers. Then all you need to show is that a number like $(45, 0, 0)$ cannot be represented in that form. This is a simple matter in, say, Cramer's rule, or just elementary Diophantine manipulation.)

Let us try for a more elegant method, as outlined above: find all possible colour combinations of the chameleons. First of all, the total number of chameleons must remain the same. This is not overly helpful in this case (although considering total populations can be a good idea sometimes in similar questions). Second, two different colour chameleons 'merge' into another colour. This merging can be focussed on. When, say, two uneven containers of water are connected at the base, the levels of water 'merge' into the middle ground. But the total amount of water remains the same. So can we say that the 'total amount of colour' remains constant?

Obviously we have to define 'total amount of colour' to make this good mathematics. Take, as an example, a grey chameleon and a crimson chameleon 'merging' into two brown chameleons. If we say, for example, that grey has a colour 'score' of 0, brown has a colour score of 1, and crimson has a colour score of 2, then the 'total colour' is preserved here.

(A 0 and a 2 join into two 1's.) But this fails when trying to merge a crimson and brown chameleon, for example. No point scoring system, it seems, can cater for all three (or even two) possibilities of merging.

The problem is due to the cyclic nature of the manoeuvres. But do not give up entirely! A partially successful (or partially failed) attempt may be a piece of a truly successful approach. (Then again, a measly amount of success is not something to be too enthusiastic about either.) Think of the primary colours, red, blue, and green. If you think of a red light beam coinciding with a green light beam, we get a doubly bright purple beam, that is, an anti-blue beam. The primary colours are cyclic, too. Can we somehow use this coloured light analogy to our advantage?

Well, the only essential difference is that in light, red and green combine into anti-blue, not blue. But wait! We can make blue equal to anti-blue by a modular arithmetic approach. With this in mind, we can try looking at our vectors (mod 2): our vector starts at $(1, 1, 1)$ and we have to stop it from going to $(1, 0, 0)$, $(0, 1, 0)$, or $(0, 0, 1)$. Unfortunately, this does not work. But now the genie is out of the bottle: we can try other moduli. The modulus (mod 3) comes quickly to mind (after all, there are three cyclic colours). Now we can try either of our two tactics to conquer this problem:

- (Vector approach) Our initial vector, $(13, 15, 17)$ is now $(1, 0, 2)$ (mod 3), and investigation shows that the exchanging of colours can only lead to the vectors $(1, 0, 2)$, $(0, 1, 2)$, and $(1, 2, 0)$, and it can never lead to any of our three objects $(45, 0, 0)$, $(0, 45, 0)$, $(0, 0, 45)$, which are all equal to $(0, 0, 0)$ (mod 3).
- (Total colour approach) Our old method of calculating 'total amount of colour' was to assign a point score to each number. Now that we know about moduli, why not use a modulus point score? Say grey has a score of 0 (mod 3), brown a score of 1 (mod 3), and crimson a score of 2 (mod 3). This works: the total point score must remain constant (because none of the three merging possibilities will change the total point score—try it yourself). The total point score initially is $13 \times 0 + 15 \times 1 + 17 \times 2 = 1$ (mod 3), while the point scores for our three objectives (45 grey, 45 brown, or 45 crimson) are all 0 (mod 3).

EXERCISE 6.1. Six musicians gathered at a music festival. At each concert some musicians played in the concerts while the others listened, as part of the audience. What is the least number of concerts needed to be scheduled in order that each musician may listen, as part of the audience, to every other musician? (Hint: Obviously not everyone can listen to everyone else in one concert, so more than one concert is needed to exhaust all the 'listening possibilities' ... think upon these lines and also of 'point-scoring' ideas, and you will get a reasonable lower bound for the number of concerts needed. Then find an example satisfying this lower bound—and you have solved it.)

EXERCISE 6.2. Three grasshoppers are on a line. Each second, one (and only one) grasshopper hops over another. Prove that after 1985 s, the grasshoppers cannot be in their starting positions.

EXERCISE 6.3. Suppose four checkerboard pieces are arranged in a square of sidelength one. Now suppose that you are allowed to make an unlimited amount of moves, where in each move one takes one of the checkerboard pieces and jumps over it, so that the new location of that piece is the same distance from the piece jumped over as the original location (but in the opposite direction, of course). There is no limit as to how far two checkerboard pieces can be in order for one to jump over the other. Is it possible to move these pieces so that they are now arranged in a square of sidelength two? (There is a particularly elegant solution to this problem, if you just think about it the right way.)

PROBLEM 6.2 (*). Alice, Betty, and Carol took the same series of examinations. For each examination there was one mark of x, one mark of y, and one mark of z, where x, y, z are distinct positive integers. After all the examinations, Alice had a total score of 20, Betty a total score of 10, and Carol a total score of 9. If Betty was placed first in Algebra, who was placed second in Geometry?

There is precious little information in this question: it seems that we know little more than the final scores. And how can one determine partial scores from the total score? But we may, because we have other data at our disposal. First of all, each time there was an exam (and we do not know yet how many exams there were) one girl scored an x, one girl scored a y, and one girl scored a z. This is an unusual piece of data. How can we exploit it?

First of all, we can try to match it with our third piece of data, that Betty was first in Algebra. Well, that means that Betty scored the highest of the three choices $x, y,$ and z. To make life easier, say that x is the biggest and z is the smallest: that is, $x > y > z$ (remembering that $x, y,$ and z are known to be distinct). We do not lose much, but we gain some simplicity: we can say that Betty scored x points in Algebra.

But for the other exam(s), we still do not know much about the various possibilities. For example, in Geometry, Alice could score z, Betty x, and Carol y, or perhaps Alice scores x, Betty y, and Carol z. Does anything stay fixed amid all these possibilities? Well—the total score per exam remains the same. No matter how the x, y, and z scores are handed out, the total score per exam must always be $x + y + z$. Do we know anything more about these total scores? Well, we know that the total score for *all* exams is $20 + 10 + 9 = 39$, we get

$$N(x + y + z) = 39,$$

where N is the number of exams. Now we have a formula containing the number of exams, which we knew little about beforehand. This should prove useful.

But one lone equation does not seem enough. However, we must keep in mind that N, x, y, z are positive integers, not just real numbers. Also, we have a fourth piece of data: x, y, and z are distinct. These weapons will reduce the possibilities of the above equation.

Well, we know that N, x, y, and z are positive integers, so the above equation is of the form

$$\text{(positive integer)} \times \text{(positive integer)} = 39.$$

So N and $x + y + z$ must be factors of 39. But the only factors of 39 are 1, 3, 13, and 39. So we have four possibilities:

(a) $N = 1$ and $x + y + z = 39$.
(b) $N = 3$ and $x + y + z = 13$.
(c) $N = 13$ and $x + y + z = 3$.
(d) $N = 39$ and $x + y + z = 1$.

But not all these possibilities hold water. Possibility (a), for instance, states that there was only one exam taken. This contradicts the semantics of the question, which implies that there are at least two exams (Algebra and Geometry). And possiblities (c) and (d), apart from seeming like a suicidal number of exams, cannot work if x, y, z are to be distinct, positive integers (that forces $x + y + z$ to be at least six). So the only possibility that has not been eliminated (c); thus there must have been three exams, and $x + y + z = 13$.

Now we have far fewer possibilities. But we still do not know two things that should be important: we do not know the exact values of x, y, and z; and we do not know how everyone scored in each exam. The first question is partially handled by the fact that x, y, and z are distinct positive integers that add up to 13, while the second question is partially answered by the fact that we know Betty scored a x in Algebra. How can we improve on these partial results?

Well, the one piece of data that has not been used fully is the individual total scores. Looking at those scores, we see that Alice did rather better than Betty and Carol, implying that she probably got high marks (i.e. x's and y's) in each subject. But Betty was first in one subject, so Alice could not have got straight x's. At best she could have got 2 x's and a y. Likewise, Carol would be unlikely to score the top mark of x in any of the exams, and it is more than likely that she would be scoring mostly z's. Can we put this speculation into solid mathematics?

The answer is at first, a 'maybe'. Let us take Alice's scores for example. At best, she can score $2x + y$. Perhaps we can prove that she scores *exactly* $2x + y$; after all, Alice scores a good deal more than any other girl; 20 is much larger than 10 or 9. What are the other possibilities for Alice's scoring? They are $2x + z$, $x + 2y$, $x + y + z$, $x + 2z$, $3y$, $2y + z$, $y + 2z$, and $3z$. The last few in the list seem too low-scoring to possibly reach 20: hopefully, they could be eliminated. But to try to prove this rigorously, we need some decent upper bounds on x, y, and z. So this is our next task: to limit x, y, and z so we can eliminate several possibilities.

All we know is that x, y, z are integers, $x > y > z$ and that $x+y+z = 13$. But this is enough to put quite good bounds on x, y, and z. Let us tackle z, for example. z cannot go too high, because then x and y will have to go high as well, and then $x + y + z$ will probably be forced to go higher than 13. To be specific: y is at least $z + 1$, and x is at least $z + 2$, so

$$13 = x + y + z \geq (z + 2) + (z + 1) + z = 3z + 3,$$

which forces $z \leq 3$. Now this bound $z \leq 3$ is the best one can do without any further information, for there is the combination $x = 6, y = 4, z = 3$.

Now let us try to do y. We can do something similar to the above, bounding x by $y + 1$. But all we can say about z is that it has a lower bound of 1. But this is enough: we get

$$13 = x + y + z \geq (y + 1) + y + 1 = 2y + 2,$$

so $y \leq 5$. Again, this is the best possible: consider $x = 7, y = 5, z = 1$. Finally, we can put a bound on x: z is at least 1 and y is at least 2, so $13 = x + y + z \geq x + 2 + 1$, so $x \leq 10$. And this is the best possible: for we have $x = 10, y = 2, z = 1$.

So we know this: $z \leq 3$, $y \leq 5$, and $x \leq 10$. But we can do even better. Remember that Betty scored an x and two other scores. Since Betty only scored 10, we know that x cannot be as high as 10. This would mean that Betty would have scored nothing for her other two exams, which is impossible: we know that all the scores are positive integers. In fact, x cannot be as high as 9 either, for Betty would have only coughed up 1 mark for the other two exams: this would mean that one exam scored nothing, again a contradiction. So we have in fact $x \le 8$. Now we can do some serious eliminating: in fact it is easily seen that the only possibility for Alice's scores is $2x + y$: all the other scoring possibilities cannot possibly reach 20. For example, $2x + z$ is at most $2 \times 8 + 3 = 19$.

So Alice scored two x's and a y. Since Betty scored an x in Algebra, Alice must have scored a y here. We can place this, together with the other information we know, into a table:

Exam	Alice	Betty	Carol	Total
Algebra	y	x	?	13
Geometry	x	?	?	13
Other	x	?	?	13
Total	20	10	9	39

We can now see that Carol must have scored a z in Algebra, as it is the only mark remaining.

We are getting closer to our goal; we know that it is either Betty or Carol that scored the second-place mark of y in Geometry. But we are still not done yet. Looking at Alice's column of the table, we have another piece of information, namely that $y + x + x = 20$. Recalling that $x > y$ and $x \le 8$, this gives us only two solutions: $x = 8, y = 4$ or $x = 7, y = 6$. But since $x + y + z = 13$, we can not have $x = 7$ and $y = 6$, since that would force $z = 0$. Thus we can only have $x = 8, y = 4$, which forces $z = 1$. So we have made a major breakthrough by solving for x, y, z completely. We can now update our table:

Exam	Alice	Betty	Carol	Total
Algebra	4	8	1	13
Geometry	8	?	?	13
Other	8	?	?	13
Total	20	10	9	39

And now it is easy to see that Betty had to have scored $z = 1$ in both Geometry and the other exam, and Carol had to have scored $y = 4$ in those exams. So the answer was that Carol scored second in Geometry.

6 : Sundry examples

> PROBLEM 6.3 (Taylor 1989, p. 16, Q.3). Two people play a game with a bar of chocolate made of 60 pieces, in a 6 × 10 rectangle. The first person breaks off a part of the chocolate bar along the grooves dividing the pieces, and discards (or eats) the part he broke off. Then the second person breaks off a part of the remaining part and discards her part. The game continues until one piece is left. The winner is the one who leaves the other with the single piece (i.e. is the last to move). Which person has a perfect winning strategy?

By the way, it is easily shown that any finite game of skill must have a winning (or drawing) strategy for one of the players. This is done by induction on the maximum length of the game. Even chess has this restriction, although no-one has found the strategy, which most believe is extremely complicated. Since there are no draws in ths game, one player must have a perfect winning strategy (as there are no draws). But who?

First, let us reduce the problem from chocolate to Mathematics. We can start by formalizing the process of breaking the bar. Anyone who has broken a chocolate probably knows that the only way to break a bar of chocolate is into two rectangles, not along some zig-zag or partial rectangle. Essentially we reduce the 6 × 10 bar into a smaller rectangle with one of the dimensions the same (see diagram, where the dotted line is the cutting edge) that is, the chocolate breaks into a bar of equal width and less length, or equal length and less width. For instance, in the picture below, the 6 × 10 bar will be broken into a 6 × 7 bar (with the 6 × 3 piece being discarded or eaten).

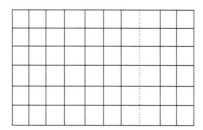

Now we need some notation for the rectangle, preferably in numbers. How do we describe a rectangle of chocolate in terms of numbers? The obvious candidate is to state the length and width of the bar. So our original bar will be a 6 × 10 bar, or perhaps (6, 10) in vector notation. The position of the chocolate is not relevant; only the size will be important. Our aim is to leave the other player at (1, 1), Our rules? We can cut off some of the width or some of the length, though not to zero or a negative number. For example, from a (6, 10) bar we can move to the following positions:

$$(6, 1), (6, 2), (6, 3), \ldots, (6, 9), (1, 10), (2, 10), \ldots, (4, 10), (5, 10).$$

In short, we can move horizontally left or vertically downward. The following diagram illustrates this abstractly, illustrating two of the possible states one can reach if one starts with the $(6, 10)$ position:

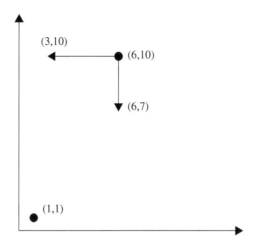

Now that we have a nice mathematical model of the chocolate, we can restate the question mathematically (but less deliciously) as

> Two players take turns moving a point on a lattice either an integer number of steps to the left or an integer number of steps downward. The point cannot pass either of the axes, and starts at $(6, 10)$. The winner is the one who reaches $(1, 1)$. Who has a perfect winning strategy?

or we could have another formulation:

> Two players take turns removing counters from two rows. Each player must remove counters from either the top row or the bottom row, but not both. At the beginning there are five counters on the top row and nine on the bottom row (this represents the point $(6, 10)$). The winner is one who takes the last counter. Who has a perfect winning strategy?

This formulation has been modified a bit by subtracting 1 from both the top and bottom row. It should give strong hints to anyone familiar with the game of Nim; those people would solve the question easily now. But we can do the question without the knowledge of Nim and related Game theory.

Now we have notation, and an abstract mathematical model. What we need now is a good grip on the problem. The problem is that the 6×10 bar has so many possibilities. We should start with a much smaller bar to experiment. Let us have a 2×3 bar to begin with.

Now the first player could leave one of the following bars behind: 1×3, 2×2, 2×1. The 1×3 bar and the 2×1 bar are foolish moves because then the second player could take everything except the last 1×1 square and win. So the first player should leave behind a 2×2 square bar. Now the second player is forced to leave behind a 1×2 or a 2×1 bar, and then the first player just breaks the remaining bar in half, leaving the 1×1 bar, and therefore winning the game. So the first player wins with a 2×3 bar.

Not much information is gained, so let us move on to another example, say a 3×3 bar. Now the first player has several options: $1 \times 3, 2 \times 3, 3 \times 2$, and 3×1. But symmetry effectively eliminates the last two choices. 1×3 is stupid, because the second player can grab everything except the last chunk and win. But 2×3 is equally bad, because we have reduced the problem to that of the last paragraph! Now the second player uses the strategy that the first player would have used in the last paragraph: get a 2×2 bar, leaving the first player with no choice but to break off a 1×2 bar, which the second player then breaks into a 1×1 bar and wins. So the first player loses with a 3×3 bar.

We solved the 3×3 problem by looking at the 2×3 problem. This suggests an induction approach for the general problem. For example, suppose we wanted to solve the 3×4 problem, and we already knew that, say, the 3×1, 3×2, 1×4, and 2×4 problems were all winners for the first player, while 3×3 was a loser for the first player. Then the strategy of the first player in the 3×4 problem would be to leave a 3×3 square with the second player, because that it is a sure loser for the second player. So the strategy for the first player is to leave the second player with bars that are sure losers for the one who has to break them. And why are these bars sure losers? Because no matter how you break them, they become sure winners for the other player. And these bars are sure winners because one can break them into a sure loser for the other player, and so on. So our strategy now is to find all the sure winners and sure losers.

1×1 is an obvious sure loser for the one who is stuck with it; it can not be broken, so the game is over. $1 \times n$ bars are all sure winners (n greater than 1), because the breaker can leave the other player with the sure loser, 1×1. Now 2×2 is a sure loser, because the breaker must end up with a 1×2 bar, which is a sure winner for the other player. Now, we can say that $2 \times n$ bars (with $n > 2$) are sure winners, because we can land the other player with the sure loser, 2×2. And so forth. We note that:

- If $a \times b$ is a sure loser, than $a \times c$ (with $c > b$) is a sure winner, because the person who breaks $a \times c$ should leave the other player with $a \times b$. For example, because we have shown that 3×3 is a sure loser, then 3×4, 3×5, 3×6, etc. are all sure winners.
- $a \times b$ is a sure loser only when all the possible moves from it are sure winners for the other player. For example, 1×4, 2×4, 3×4, and by symmetry 4×3, 4×2, 4×1 are all sure losers, as we have shown above, hence 4×4 must be a sure loser.

One can continue this systematic method, eventually reaching 6×10. But why do not we be more mathematical? There should be a pattern to the sure winners and losers. Well, what are the winners and losers we know so far? The sure winners that we worked out already are

$$
\begin{array}{cccccc}
 & 1 \times 2 & 1 \times 3 & 1 \times 4 & 1 \times 5 & \ldots \\
2 \times 1 & & 2 \times 3 & 2 \times 4 & 2 \times 5 & \ldots \\
3 \times 1 & 3 \times 2 & & 3 \times 4 & 3 \times 5 & \ldots \\
4 \times 1 & 4 \times 2 & 4 \times 3 & & 4 \times 5 & \ldots \\
\end{array}
$$

while the sure losers that we have identified are 1×1, 2×2, 3×3, and 4×4.

This is pretty convincing evidence that the only sure losers are $n \times n$ bars: that is, square bars, and all others are sure winners. Once we have this conjectured strategy, we do not even have to prove it (although you can, with induction): we just have to apply it. Remember we want to leave the opponent with losers. Once we guess what the losers are, we can make the strategy to always force them on the opponent. If the strategy works all the time, then fine. If not, then the guess was wrong. To summarize, if our guess is correct, the best strategy is to give the other player a square. So this means that for the 6×10 bar, the first player has the following strategy:

> Break the chocolate so that a 6×6 square is left (a sure loser to the second player). Then whatever the second player does, convert the bar back into a square. For example, if the second player leaves a 6×4 bar, you square it back to a 4×4 bar. Repeat this process, always leaving the other player with the square formation, until at last you leave the opponent with the 1×1 formation (making him/her lose).

This strategy actually works, because whenever the opponent breaks the square, he gets a non-square that can be easily converted to a square again. And because the size of the chocolate is decreasing, the square conversion must eventually lead to a 1×1 square. And so with a bit of semi-rigorous maths we have ended up with a working strategy, which is what we were after.

Anyway, this is a standard approach in solving games of skill: determine all winning and losing positions, then always move to a winning position. Any decent skill-game player uses this method, except that they have an imprecise idea of winning and losing positions, only 'favourable' and 'unfavourable' positions. After all, do not we say a move in, say, chess, is a 'good' or a 'bad' game because it creates a favourable or unfavourable position? Few players of chess succeed by moving randomly, not trying to improve their position.

EXERCISE 6.4. Two players play a game starting with 153 counters. Taking turns, each player must remove between one and nine counters from the game. The person who removes the last counter wins. Does either the first player or second player have a guaranteed winning strategy, and if so, what is it?

EXERCISE 6.5. Two players play a game with n counters. Taking turns, each player must remove a number of counters which is a power of d. The person who removes the last counter wins. For the following values of d, determine for which values of n the first player has a winning strategy, and for which values of n the second player wins.

(a) $d = 2$.
(b) $d = 3$.
(c) (*) $d = 4$.
(d) (*) The general case.

EXERCISE 6.6. Repeat the previous exercise, but now the objective is to lose, that is, to force the other player to take the last counter. (If one happens to be thinking in the right way, the answer falls out easily.)

EXERCISE 6.7. Imagine a three-dimensional version of Problem 6.3 in which one starts with a $3 \times 6 \times 10$ bar and can break the bar in any of the three dimensions. Which player wins, and what is the winning strategy?

EXERCISE 6.8 (**). In the game of *Gomoku*, the two players (White and Black) take turns placing a stone of their colour on a 19×19 board. A player wins if he or she manages to obtain five stones in a row (in any orientation); if all the squares on the board are filled without a five-in-a-row, the game is a draw. Show that the first player has a strategy which guarantees at least a draw. (Hint: you have to argue by contradiction. Show that if the first player cannot force at least a draw, then the second player has a winning strategy. Now make the first player 'steal' that strategy.)

PROBLEM 6.4 (Shklarsky *et al.* 1962, p. 9). Two brothers sold a herd of sheep. Each sheep sold for as many rubles as the number of sheep originally in the herd. The money was then divided in the following manner. First the older brother took 10 rubles, then the younger brother took 10 rubles, then the older brother took another 10 rubles, and so on. At the end of the division the younger brother, whose turn it was, found that there were fewer than 10 roubles left, so he took what remained. To make the division fair, the older brother gave the younger his penknife, which was worth an integer number of roubles. How much was the penknife worth?

The first reaction should be that this question seems to have not enough information. Second, the question does not seem to be rigorous enough. But it is wrong to give up hope before any attempt has been made to solve it; take a look at Problem 6.2, which had even less information to begin with but could still be solved.

We should start by trying to formulate the problem in terms of equation. For this we need some variables. First of all, we notice that the price of the penknife is ultimately dependent on the number of sheep, which is the only independent variable here. (that is, knowing the number of sheep determines everything.) Let us suppose there were s sheep. Then they all sold for s rubles, so the total windfall is s^2 rubles.

Now we have to see how this division system works. Suppose the number of rubles was 64. Then the older brother took 10, then the younger took 10, and so on. It transpires that the last four roubles go to the older brother, not the younger, so the problem cannot work here. Remember that part of the data given is the fact that the younger brother was the last to get the cash. How can we say this mathematically?

To do this mathematically we need lots of equations and variables (enough equations to describe the situation but not enough to introduce confusion and superfluousness). Suppose that the younger brother already took n lots of 10 roubles before he was shortchanged. Then the older brother also took n lots of 10 roubles, plus another 10-rouble pack just before the younger brother came up short, say having only a roubles remaining (a being something between 1 and 9 inclusive; the wording of the problem seems to suggest that a is non-zero). So the total number of roubles had to be

$$s^2 = 10n + 10 + 10n + a$$

or

$$s^2 = 10(2n+1) + a.$$

But what has this to do with the penknife? The dependent variable which we want to solve is p, the price of the penknife. We need an equation connecting p with something else, preferably s, which is the independent variable. Now before the exchange of the penknife the older brother had $10n + 10$ roubles and the younger brother had $10n + a$. Once the exchange of the penknife occurred the older brother had a profit of $10n + 10 - p$ and the younger brother had a profit of $10n + a + p$. For the exchange to be fair, these two profits have to be equal. Equating them leads eventually to the useful equation connecting p to a:

$$a = 10 - 2p. \tag{1}$$

Now we can plug back into an older equation and get an equation connecting p and the other variables. We get (cancelling a in the process)

$$s^2 = 20(n + 1) - 2p. \tag{2}$$

We have to somehow use these equations to solve for p. It looks like there is not enough information here, because we are not given what s, n, or a are. How do we narrow things down further? The root problem is that there are too many unknowns floating around here. We can eliminate some of them by modular arithmetic. For instance, in (2) we can take (mod 20) to eliminate the n, obtaining

$$s^2 = -2p \pmod{20}.$$

This is getting us closer to our goal of working out p, but we still have this pesky s to deal with. Fortunately, we can capitalize on the fact that squares have a restricted choice of values in modular arithmetic. In fact, in (mod 20), the squares must take a value of 0, 1, 4, 5, 9, or 16. In other words, we have

$$-2p = 0, 1, 4, 5, 9, \text{ or } 16 \pmod{20}$$

And solving for p (and remembering that $2p$ has to be even) we have

$$p = 0, 2, 8 \pmod{10}.$$

So we have an equation concerning p, but we have not been able to pin it exactly. All this says that the penknife could be worth 0 rubles, 2 rubles, 8 rubles, 10 rubles, 12 rubles, ... But the penknife could not be too expensive, could it? after all, the younger brother only missed out on 10 rubles or less... thinking upon these lines eventually allows you to remember that p is not only connected to n and s, it is connected to a, and a is restricted to between 1 and 9. Recalling (1), this means that $0 < p < 5$, which, when coupled with the other equation concerning p, nails the price of the penknife down to 2 rubles. (Note that this argument works even if we allow a to be zero.)

Curiously, while there is enough information to determine the price of the penknife, there is not enough information to determine the price or number of the sheep. In fact, all we can say about s is that $s = \pm 4 \pmod{20}$; thus the number of sheep could be 4,16,24,36,44,56, ...

With puzzles like these, you need all the information you can get. The best way is to spread out all the information in the puzzle and write each piece separately, for example, like so:

(a) there was a square number of roubles divided;
(b) the younger brother missed out a piece of his share;
(c) the piece missing had to be balanced by the penknife.

One should then reduce these facts to equations as quickly as possible:

(a) $s^2 = 10(2n + 1) + a$;
(b) $0 < a < 10$;
(c) $a = 10 - 2p$.

One should try to capture each piece of information, no matter how useless-looking. For example, I could have put in that n was probably non-negative, or than p probably had to be positive (why mention a worthless penknife in the question), that there were a positive integer number of sheep, and so on. Once everything is sealed into equations things become a lot easier to manipulate correctly.

References

> Books, like friends, should be few and well-chosen.
> Samuel Paterson, Joineriana

AMOC (Australian Mathematical Olympiad Committee) Correspondence Programme (1986–1987), Set 1 questions.

Australian Mathematics Competition (1984), *Mathematical Olympiads: The 1984 Australian Scene*, Canberra College of Advanced Education, Belconnen, ACT.

Australian Mathematics Competition (1987), *Mathematical Olympiads: The 1987 Australian Scene*, Canberra College of Advanced Education, Belconnen, ACT.

Borchardt, W.G. (1961), *A Sound Course in Mechanics*, Rivingston, London.

Greitzer, S.L. (1978), *International Mathematical Olympiads 1959–1977* (New Mathematical Library 27), Mathematical Association of America, Washington, DC.

Hajós, G., Neukomm, G., and Surányi, J. (eds) (1963), *Hungarian Problem Book I, based on the Eötvös Competitions 1894–1905*, (New Mathematical Library 11), orig. comp. J. Kürschák, tr. E. Rapaport, Mathematical Association of America, Washington, DC.

Hardy, G.A. (1975), *A course of Pure Mathematics*, 10th eds., Cambridge University Press, Cambridge.

Polya, G. (1957), *How to solve it*, 2nd ed, Princeton University, Princeton.

Shklarsky, D.O., Chentzov, N.N., and Yaglom, I.M. (1962), *The USSR Olympiad Problem Book: Selected Problems and Theorems of Elementary Mathematics*, revd. and ed. I. Sussmar, tr. J. Maykovich, W.H. Freeman and Company, San Francisco, CA.

Taylor, P.J. (1989), *International Mathematics: Tournament of the Towns, Questions, and Solutions, Tournaments 6 to 10 (1984 to 1988)*, Australian Mathematics Foundation Ltd, Belconnen, ACT.

Thomas, G.B. and Finney, R.L. (1988), *Calculus and Analytic Geometry*, Addison-Wesley, Reading, MA.

Index

2, powers of 14–18
9, multiples of 9, 11–13
18, multiples of 12–13

algebra 35
  examination marks problem 86–9
  polynomials 41–7
analysis of functions 36–40
analytic geometry
  line segments 77–9
  partitioning of rectangles 74–7
  square swimming pool problem 79–82
  vector arithmetic 69–74
angles
  in circles 50–1
  notation 3, 51
  proof of equality 58, 66–8
  of triangles 50–4
anti-symmetry 25, 26, 30
areas of triangles 58
arithmetic progression, lengths of triangle 1, 2–7

Bernoulli polynomials 24

chains, partitioning of rectangle 76–7
chameleon colour combinations 83–5
chessboard problem 84
chocolate-breaking game 90–3
chords, subtended angles 51, 67
circles, angles in 50–1
circle theorems 49–50, 57, 67
colour combinations, chameleons 83–5
concurrence, perpendicular bisectors
  of triangle ix
conjectures 4
consequences, proof of 4
constant polynomials 42
constructions 58–61
contradiction, proof by 65–6, 76–7
coordinate geometry 55, 58
  use in constructions 59
coprime numbers 10
cosine rule 3, 53
cubes, sum of 35
cubic polynomials 42

cyclic manoeuvres 85
cyclic quadrilaterals 58

data
  omitting it from problem 5
  recording it 3
  understanding it 2
degree of a polynomial (n) 41, 42
degrees of freedom, polynomials 46
diagonals, lengths of 69–74
diagrams 3, 4
diameter, angle subtended by (Thales' theorem)
  49–50, 57, 67
digits
  rearrangement 14–19
  summing 10–14
  powers of 2 16–18
Diophantine equations 19–22
direct (forward) approach 54–5
divisibility
  sums of powers 23–6
  sums of reciprocals 27–33
division system, price of penknife 95–7

elegance of solutions ix
elimination of variables 61
equations, use of 3–4
equilateral triangles, construction 58–60
Euclidean geometry 49–50
  angles in circles 50–1
  angles of triangles 51–4
  constructions 58–61
  equating angles 66–8
  ratios 55–8
  squares and rectangles 62–6
'Evaluate…' problems 1–2
examination marks problem 86–9
exponent variables 20–2

factorization techniques, Diophantine
  equations 21
factors 87
factors of polynomials 42, 44, 45–7
  roots of 43
Fermat's last theorem 20
'Find a…'/Find all…' problems 1, 2

Index

finite problems, simplification 13
formulae, use of 3–4
forward (direct) approach 54–5
functions
  analysis 36–40
  polynomials 41–7

generalization 4, 15
geometry *see* analytic geometry; Euclidean geometry
guessing 14, 80–2

Heron's formula 3, 4, 6–7
homogeneous polynomials 42

induction approach 92
induction, proof by 37–8, 40
inequalities 88
  in analysis of functions 36–40
  in Euclidean geometry 65–6
infinite products 28
integer lengths, rectangles 74–7
irreducible polynomials 42
isosceles triangles 52
'Is there a...' questions 1, 2

Lagrange's theorem 9
levels of difficulty viii
line segments, analytic geometry 77–9

matrix algebra 35
modification of problems 4–5
(mod n) notation 10
modular arithmetic 9, 10
  Diophantine equations 21–2
  powers of 2 17–18
  squares 96
  sums of powers 23, 24–6
  sums of reciprocals 28–33
  vectors 85
multiples of 9 9, 11–13

natural numbers 10
Nim 91
notation 3
  vectors 84, 90
number theory 9–10
  digit rearrangement 14–19
  digit summing 10–14, 16
  Diophantine equations 19–22
  sums of powers 23–6
  sums of reciprocals 27–33
  numerators, reduced 27–8

objectives of problems 2

p-adics 9
pairwise cancelling 25, 26, 30, 32
parallel lines 58–60
parameterization 44
partitioning of rectangle 74–7
periodicity 23–4
physical constraints 3
'pocket mathematics' 36
polygons, lengths of line segments 69–74
polynomials 41–3
  factorization 44, 45–7
  and reciprocals 43–4
powers of 2
  digit rearrangement 14–19
  digit sums 16–18
powers, sums of 23–6
prime numbers 10
problem types 1–2
proof by contradiction 65–6, 76–7
proof by induction 37–8
  strong induction 40
proving results 6
pseudo-coordinate geometry 58
Pythagoras' theorem 57

quadratic formula 20, 42
quadratic polynomials 42
quadrilaterals, midpoints of sides 50

ratios, Euclidean geometry 55–8
rearrangement of digits 14–19
reciprocals
  and polynomials 43–4
  sums of 27–33
rectangles
  chocolate-breaking game 90–3
  partitioning 74–7
  in a square 62–6
reduced numerators 27–8
reformulation of problems 4
representation of data and objectives 3
reversal of problems 5
roots of factors of polynomials 43
roots of polynomials 42, 44, 46
rotations 60

'Show that...' problems 1–2
similar triangles 56, 57
simplification of problems 4, 5, 6, 13, 19, 91–2
sine rule 3, 53, 54
skill games 93
special cases 4–5
  in geometry problems 58
square roots 20
squares, modular arithmetic 96
squares (Euclidean geometry) 62–6
square swimming pool problem 79–82
Steiner's theorem of parallel axes 74

steps in problem-solving 1
strong induction 40
sums of cubes 35
sums of digits 10–14
sums of lengths of line segments 79
sums of powers 23–6
  powers of 2 16–18
sums of reciprocals 27–33
symmetry 30, 32, 73–4

tables 89
tangents to circle 57
Thales' theorem 49–50, 57, 67
triangle inequality 3, 79
triangles
  angles of 50–4
  areas of 58
  concurrence of perpendicular bisectors ix
  lengths in arithmetic progression 1, 2–7
  similar 56, 57
trigonometry 53–4
trivial polynomials 42

variables 3
  elimination of 61
  exponent 20–2
vector geometry 55, 69, 73–4
vectors
  chameleon colour combinations 84–5
  chocolate-breaking game 90

Wilson's theorem 9

Lightning Source UK Ltd.
Milton Keynes UK
UKHW030214140121
377021UK00001B/1